Lecture Notes in Earth Sciences   105

Editors:
S. Bhattacharji, Brooklyn
H. J. Neugebauer, Bonn
J. Reitner, Göttingen
K. Stüwe, Graz

Founding Editors:
G. M. Friedman, Brooklyn and Troy
A. Seilacher, Tübingen and Yale

Friedemann Wenzel (Ed.)

Perspectives
in Modern Seismology

 Springer

Editor

Friedemann Wenzel
Universität Karlsruhe
Geophysikalisches Institut
Hertzstr. 16
76187 Karlsruhe, Germany
E-mail: Friedemann.Wenzel@gpi.uni-karlsruhe.de

Library of Congress Control Number: 2004114744

"For all Lecture Notes in Earth Sciences published till now please see final pages of the book"

ISSN 0930-0317
ISBN 3-540-23712-7 Springer Berlin Heidelberg New York

Springer is a part of Springer Science+Business Media

springeronline.com

© Springer-Verlag Berlin Heidelberg 2005
Printed in Germany

Cover design: Erich Kirchner, Heidelberg
Typesetting: Camera ready by author
Printed on acid-free paper   32/3142/du - 5 4 3 2 1 0

Contents

Introduction

Friedemann Wenzel

Geophysikalisches Institut, Universität Karlsruhe, Hertzstr. 16, D-76187 Karlsruhe, Email: friedemann.wenzel@gpi.uni-karlsruhe.de

This volume reflects interdisciplinary research in international cooperation at the Geophysical Institute of Karlsruhe University. A considerable part is contributed from the unique cooperation of Earth Sciences and Civil Engineering in the field of strong earthquakes in the Vrancea region of Romania and its capital Bucharest (Part 1). Part 2 is reviewing the results of deep seismic tomography from mantle plumes, deep lithospheric properties from Russian Nuclear explosion data and high-resolution imaging in applied seismics. Part 3 is dealing with the effect of tectonics on plate motions and its effects on civilisation.

The first part of this book reflects efforts of the geoscience and the civil engineering community in earthquake mitigation. Earthquake early warning systems (EWS) as components of earthquake information systems can be utilised by a number of 'customers' ranging from chemical plants to transportation services. The paper by F. Wenzel discusses the most recent developments in this field and highlights the specific future of the EWS for the Romanian capital Bucharest. Another result of collaboration between civil engineers and geoscientists is presented in Fäcke et al. who study the seismic safety of long-span bridges crossing the river Rhine. Key questions refer to the appropriate level of ground motion that has to be expected but also to the construction type. The authors reveal specific failure mechanism for a box girder bridge. Ground motion modelling is often done with empirical laws adopted from several areas without consideration of information on the geology of the earthquake source and the geology of the medium in which the source-generated waves propagate. Gottschämmer et al. demonstrate that these shortcomings can be surpassed if modern tools, such as three-dimensional modelling techniques for elastic wave propagation are used. In this case the strike-slip sources and the sedimentary in-fill of the Dead Sea Rift cause considerable wave-amplification in the rift, which must be taken into account in hazard assessment. The role of nonlinearity in seismology is an on-going debate that can be resolved only if seismologists, geotechnical engineers and geologists join efforts and develop a consistent picture of seismological strong motion observations, geotechnical laboratory experiments and physical models of nonlinear wave propagation. Marmueranu et al. make an attempt in this direction with emphasis on observations of strong Vrancea seismicity in Romania.

The second part of the book presents examples of seismic exploration techniques that range from ore and hydrocarbon exploration problems to the deeper mantle of the earth. J. Ritter studies mantle upwellings with passive seismology on continents in the Massif Central and more recently in the German Eifel region. A major challenge for geoscientists in the next decades is to identify more of these small-scale plumes and provide models that are based on observations. Complementary fluid- and geodynamic simulations and petrophysical modelling is necessary to fully understand the dynamics of small-scale mantle plumes as well as their contribution to the mixing of mantle material.

Ryberg et al. provide an overview on two decades of active upper mantle investigations in Northern Eurasia that revealed features previously hidden to passive seismology. Among them is clear evidence for a 520 km discontinuity and a fine-structure of the sub-lithospheric mantle that allows for efficient high-frequency wave propagation. Jäger et al. demonstrate the huge progress made in seismic imaging of hydrocarbon reservoirs. By applying suitable weight functions in the migration process the geometrical spreading effects of propagating waves can be compensated. Such an approach is called 'true-amplitude migration'. In this way, the output amplitudes are related to the reflection coefficient. As a consequence, detailed amplitude versus offset (AVO) or angle (AVA) analysis may be performed and, thus, the search for, e.g., reservoirs is improved.

Duveneck et al. present a novel migration technique that utilises all 3 recorded components of Vertical Seismic Profiling (VSP) data. The Polarisation Migration is successfully tested on synthetic elastic Finite Difference seismograms of the pure scattering response of a complex object in a simple VSP geometry. A VSP dataset measured near a known orebody is processed and migrated using the described Polarisation Migration. The migration result shows a clear anomaly at the actual position of the orebody.

The third part of this book deals with tectonics, on a plate scale, with modern approaches to modelling plate tectonics, and simultaneously understanding the implications of tectonics on civilisation. Ben-Avraham et al. argue that the present physiography of the Dead Sea Rift developed slightly before man started his way out of Africa. This can be understood as one example where tectonic processes controlled the paths of human migration and development. The Dead Sea Rift is part of the African-Arabian-European collision scenario. D. Heidbach presents results of large-scale mechanical modelling of this collision with a Finite Element technique. The Aegean-Anatolian region belongs to the areas with the highest seismicity on a global scale. It influenced human societies since thousands of years and unfortunately even today.

Sperner et al. discuss the evidence that by looking at the Romanian Vrancea earthquake region as the site of a process in the deep earth, where a previously subducted oceanic slab segment is currently detached and will be returned to the deeper earth's mantle. This modern view on the recent tectonics of SE-Romania is a result of interdisciplinary research with the Collaborative Research Center 461 'Strong Earthquakes: A Challenge for Geosciences and Civil Engineering' during the past years. Ismail-Zadeh et al. provides quantitative modeling of this geodynamic process that allows a physical understanding of the detachment process.

The papers of this volume have been presented during a symposium in honour of Professor Karl Fuchs' 70th birthday. Karl Fuchs was born in 1932 in Stettin. He studied Geophysics in Hamburg, London, and Clausthal. After two years of hydrocarbon exploration in Algeria and Brazil he returned to Clausthal, where he finalised his Ph.D. in 1963. After research work on crustal structure and its effect on seismic spectra in St. Louis and Dallas he returned to Germany, where he joined Stephan Müller in 1965, who had established the Geophysical Institute of Karlsruhe University in 1964. In 1971 Karl Fuchs became a professor of Geophysics and director of the Geophysical Institute. Within the next 30 years he developed the institute to an international key player in lithospheric geophysics. He became an emeritus in 1999 but continues research and supports geophysics until today. Karl Fuchs supervised 150 diploma students and 70 Ph.D. theses. Karl Fuchs initiated a number of international research initiatives. Between 1981 and 1995 he was head of the Collaborative Research Center 'Stress and Stress Release in the Lithosphere'. The World Stress Map (WMS) task force of ILP was a result of this project. Since 1995 the WSM is run by the Heidelberger Academy of Sciences and Humanities with Karl Fuchs as Principal Investigator. He initiated EUROPROBE that was sponsored by the European Science Foundation and bridged the gap between eastern and western Europe when the iron curtain fell. He played a key role in setting up the International Continental Drilling Program during his time as President of the International Lithosphere Program (ILP) between 1985-1990.

Karl Fuchs is a member of the Heidelberg Academy, an honorary member of the American Geophysical Union, the American Association of the

Advancement of Science (AAAS), the British Royal Astronomical Society (RAS), and the Deutsche Geophysikalische Gesellschaft (DGG). He is Honorary Professor of Bucharest University and a member of the Academia Europea. In 2002 he received the Heitfeld-Prize by the Alfred-Wegener Foundation for his outstanding contributions to propagation of seismic waves and international cooperation in the Earth sciences.

An Earthquake Early Warning System for the Romanian Capital

Friedemann Wenzel

Geophysical Institute, Karlsruhe University, Hertzstr. 16, 76187 Karlsruhe, Germany, Email: friedemann.wenzel@gpi.uni-karlsruhe.de

Abstract

Although earthquake early warning systems have inherently short warning times between seconds and a minute they become increasingly important as part of real time information systems that can be designed to support disaster management in earthquake prone areas. In some cases (Mexico City, Bucharest) the hazardous earthquakes are bound to a rather distant tectonic unit so that a fairly large warning time can be achieved. This is demonstrated for the case of the Romanian capital Bucharest, where the warning time could be around half a minute.

1. Introduction

The discussion on early warning systems (EWS) was intensified during the final year of the International Decade of Natural Disaster reduction (IDNDR) and is well documented in Zschau and Küppers (2003). Under the restructured international effort for disaster mitigation, the International Strategy for Disaster Reduction (ISRD, http://www.unisdr.org/), early warning plays again a key role, with a broad view within the international community that

1. the technical feasibility for early warning has significantly increased during the past years
2. the concept of early warning should be based on a broader concept, that puts it right in the centre of disaster management.

So far the focus of the discussion is on disaster types that allow inherently long warning times such as hurricanes (days), volcanoes (days to hours), tsunamis (hours to 20 minutes). However, a new paradigm in understanding early warning systems which extends beyond the purely technical issues provides new opportunities for systems with inherently short warning times.

In the case of earthquakes, warning times are fairly small, ranging from seconds to a maximum of about one minute for Mexico City. However, even this small time window can provide opportunities to automatically trigger measures, such as the shutdown of computers, the rerouting of electrical power; the shutdown of disk drives, the shutdown of high precision facilities, the shutdown of airport operations, the shutdown of manufacturing facilities, the stoppage of trains, the shutdown of high energy facilities, the shutdown of gas distribution, the alerting of hospital operating rooms, the opening of fire station doors, the starting of emergency generators, the stoppage of elevators in a safe position, the shutoff of oil pipelines, the issuing of audio alarms, the shutdown of refineries, the shutdown of nuclear power plants, the shutoff of water pipelines, and the change to a safe state in nuclear facilities (Harben, 1991).

Some of these measures have been implemented or are under consideration in Japan, Mexico, Taiwan, California, Romania, Turkey and other locations. This paper explains the physical basis of EWS' and necessary system components, discusses several examples with emphasis on an EWS for the Romanian capital of Bucharest, provides the basis for a cost-benefit analysis and puts earthquake EWS' in the broader context of real-time information systems and disaster management.

2. Examples and Design Principles

The first earthquake EWS was installed along Japanese railway lines in the mid 1950's. When very fast trains were operated (the Tokaido Shinkansen since 1964; the Tohoka Shinkhansen since 1982), improved systems were introduced. Alarm accelerometers installed every 20 km issue an alarm if the value of 40 cm/s^2 horizontal acceleration is exceeded (Nakamura, 1989). Nakamura (1996) reported that the system functioned appropriately during the Kobe (Jan. 17, 1995, $M_w = 6.9$) earthquake. The original system focusing on railways developed into an Urgent Earthquake Detection and Alarm System (UrEDAS, Noda and Meguro, 1995) that issues alarms after initial ground shaking occurs. It is based on the detection and classification of P-wave motion.

The Taiwanese EWS utilises a specific plate tectonic scenario for its design, characterised by the Philippine plate colliding with the Eurasian plate along the East coast of Taiwan with associated strong seismicity. The EWS is based on the potential time difference between detection of the event by strong motion stations and the arrival of S-waves at large urban centres such as Taipei. For an event at a distance of 100 km and focal depth of 20 km Lee et al. (1996) estimate a warning time of about 17 seconds with the assumption that locating and determining the size of the event takes 10 seconds. The development of the system has been underway since 1992, and a prototype has been implemented based on 12 accelerometers and a broadband station located in Hualien.

The Taiwanese system contains features that are also characteristic for the Mexico EWS and the system for Bucharest:
1. The most hazardous source of seismicity is some distance away from the elements of risk. This provides a traveltime difference between P-waves recorded in the vicinity of the source and destructive S- or surface-waves arriving at the site of risk.
2. The first arriving waves (P-waves) can be used to quickly derive the information on the potential size and hazard of the event.
3. In order to estimate the expected ground motion one needs to determine hypocenter and magnitude of the event. The technology to do this in a few seconds has become available only recently. In the mid 60s magnitude and hypocenter determination required time in the order of an hour. 20 years later, utilizing modern communication and digital computing source parameter determinations took at least a minute. The best technological basis today is continuous monitoring with a real-time data stream arriving at the processing unit.

A second prototype system is meanwhile implemented in Taiwan utilizing the existing communication facilities of the Taiwan Regional Telemetered Seismic Network operated by the Central Weather Bureau. Descriptions of this system can be found in Shin et al. (1996), Teng et al. (1997) and Wu et al. (1997).

The most recent system has been installed around the Turkish megacity of Istanbul (M. Erdik, personal communication). It is called the Istanbul Earthquake Rapid Response and Early Warning System and materialized in the aftermath of the catastrophic August 17, 1999 Kocaeli earthquake with magnitude 7.6 and about 20,000 fatalities.

In California the first modern suggestion including a feasibility study was forwarded by Heaton (1985).

The most advanced EWS is the 'Seismic Alert System (SAS)' for Mexico City (Espinoza-Aranda et al., 1995). The disastrous M = 8.1 earthquake of September 19, 1985 caused the collapse of 200 high-rise buildings,

10.000 dead and 30.000 seriously injured people although the seismic source was about 350 km to the SW of the city. The 1985 event triggered design and installation of an EWS.

Twelve digital strong motion field stations are distributed along the Guerrero coast in 25 km intervals about 320 km SW of Mexico City where strong earthquakes related to the subduction of the Cocos plate beneath the North American plate occur. The large distance allows for a 60 to 75 s warning time (Espinosa-Aranda et al., 1995). A critical test with the largest event since the operational phase started (Sept. 14, 1995, M = 7.3) was successful with a warning message broadcast 72 seconds prior to arrival of strong ground motion.

On the basis of empirical relations between the magnitude of a subduction earthquake and the evolution of the accelerometer signal the magnitudes of events within about 100 km radius are estimated by each station separately, and sent to CIRES processing facility in Mexico City. Upon co-incidence of two M > 6 messages an alarm is issued via UHF to the general public and specifically to schools. The alarm issued before the M = 7.3 1995 event reached about 20% of the population. Regular drill, exercises and advertisements help to raise preparedness.

Technically speaking an EWS consists of 4 components: (1) a monitoring system composed of various sensors, (2) a real-time communication link that transmits data from the sensors to a computer, (3) a processing facility that converts data to information, and (4) a system that issues and communicates the warning. Specific seismotectonic circumstances may allow keeping some of those components fairly simple and robust.

3. Design of an EWS for Bucharest

Recently the authors of this paper started to design an earthquake EWS for the Romanian capital of Bucharest (Wenzel et al., 1999). The design relies on specific seismotectonic properties of the intermediate depth earthquakes that determine the hazard of Bucharest. As will be shown these specifics include the epicentral stationarity of the threatening earthquakes and their consistent source mechanisms. Together these features allow to design a simple, cheep and robust system that allows for 25 s warning time.

Within the last 60 years Romania has experienced 4 strong Vrancea earthquakes (Oncescu and Bonjer, 1997): Nov. 10, 1940 (M_w = 7.7, 160 km deep); March 4, 1977 (M_w = 7.5, 100 km deep); Aug. 30, 1986 (M_w = 7.2, 140 km deep); May 30, 1990 (M_w = 6.9, 80 km deep). The latter event was followed by a M_w = 6.3 aftershock on May 31, 1990. The 1977 event had catastrophic character with

35 high-risk buildings collapsed and 1500 casualties, the majority of them in Bucharest.

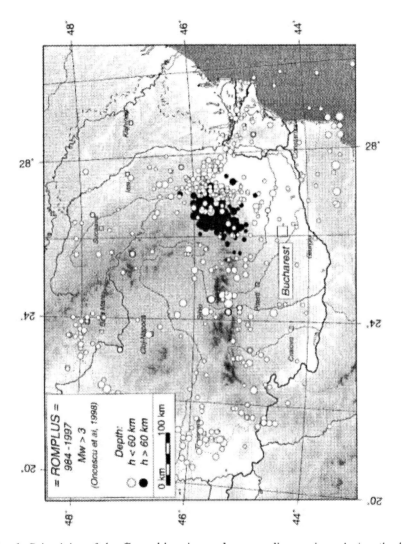

Fig. 1. Seismicity of the Carpathian Arc and surrounding regions during the last millennium. Note that the main source of earthquake hazard, the intermediate depth events beneath 60 km depth are clustered in a small area.

The entire historic catalogue (Oncescu et al., 1999) indicates that although larger crustal events occurred in the Carpathians as well as along the Shabla fault zone in Northern Bulgaria the seismic hazard of Bucharest is almost entirely controled by the intermediate depth Vrancea events.

The instrumentally well-located intermediate depth seismicity is even more localised than indicated in Figure 1. The epicenters are confined to a region of 30×70 km, with an average epicentral distance to Bucharest of about 130 km. This geometric relationship between hypocenters being confined to a small source volume and at a fixed distance to the capital allows the design of an EWS with a warning time of about half a minute for all potential intermediate deep earthquakes. A Romanian EWS would thus be similar to the Mexican case, where the site of strong earthquakes is constrained to the plate boundary at significant distance from Mexico City. For both sites a fairly constant warning time can be made available, although Bucharest can only utilise about one third of the time available to Mexico City.

The geometry of typical seismic source, epicenter and Bucharest is sketched in Figure 2, whereas Figure 3 shows P- and S-wave arrival times for a standard velocity model (Oncescu et al., 1999) and two source depths of 100 and 150 km. If Bucharest is located at an offset of 150 km from the epicenters for both source depths, the time delay between the S arrival in Bucharest and the P arrival in the epicenter is about 28 s. Obviously this represents the maximum achievable warning time.

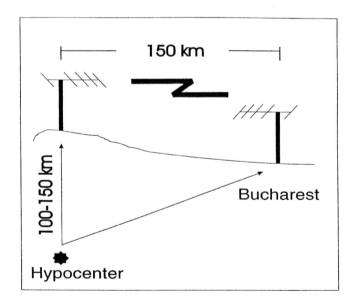

Fig. 2. Relative positions of hypocenters, epicenters and the target area (Bucharest).

Empirical observation of P- and S-arrival times for various source depths (Wenzel et al., 1999) confirm, that 25 to 30 seconds warning time represent a stable estimate for Bucharest. The expected level of ground motion in Bucharest expressed in horizontal peak-ground acceleration strongly depends on magnitude and depth of the event. Table 1 indicates the trade-off between both parameters. It shows values for horizontal PGA (related to S-wave motion) for 3 source depths and 4 magnitudes indicating that an increase of 50 km source depth can be balanced by an increase in magnitude by 0.5 units.

Table 1. Horizontal peak ground acceleration (in cm/s^2) of Vrancea earthquake predicted in Bucharest with the attenuation relation of Lungu and Coman (1994) for variable source depth (h) and moment magnitude (M_w).

M_w \ h(km)	100	150	200
7.0	116	67	39
7.5	197	114	66
8.0	335	194	113
8.5	570	329	191

Thus, the prediction of ground motion in Bucharest from source parameters requires a fairly precise determination of both magnitude and focal depth, which in turn requires operation of an extended seismic network, communication between stations and central processing facility, and data being adequately processed. With regard to an EWS it becomes a complex and vulnerable system and a lot of time would be lost for data processing before a warning could be issued.

If, on the other hand, only the P-wave amplitude of an epicentral station suffices as indicator for a strong earthquake, the system design becomes very simple and thus reliable, and precious time can be saved. The prediction of the level of ground motion the capital will experience must then be established on the basis of scaling relations between P amplitudes in the epicenter and S amplitudes in Bucharest.

The geophysical basis for the existence of this scaling relation is found in the consistent fault plane solutions of all strong and most moderate and weak Vrancea earthquakes. All events have a very similar radiation pattern. The requested scaling relation has been established experimentally with various earthquakes recorded at the epicentral station MLR and a station in Bucharest (BUC1). In order to harmonise the data from several types of instruments used in this study (SMA-1, FBA23, S13/SH-1, S13) data have to be subjected to narrow-band filtering (1-2 Hz) in order to arrive at the simple relations (Figure 3).

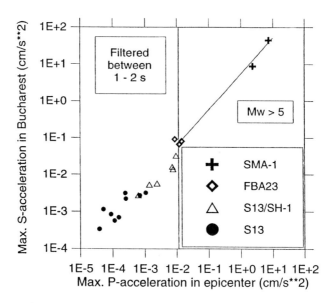

Fig. 3. Observed P and S accelerations filtered between 1-2 s: S amplitudes in the target area (S_{BUC}) are 10 times larger than P amplitudes in the epicentral area (P_{EPI}).

$$S_{BUC} \approx 10 \cdot P_{EPI} \tag{1}$$

and

$$S_{EPI} \approx 10 \cdot P_{EPI} \tag{2}$$

For details see Wenzel et al. (1999). The formulae indicate that the shear ground motion in the epicenter and in Bucharest is quite similar. This does not come unexpected as the travel distances from the intermediate deep source to both sites are also quite similar.

The purpose of any EWS is to issue messages at sites of interest before the destructive seismic energy arrives. This requires 4 components. A strong motion detector system, processing unit, a communication system transmitting the messages produced in the processing stations and a user receiver system. In contrast to sites where potential earthquakes can be distributed over a large area (e.g. California; Harben, 1991) or along an extended coastline (e.g. Mexico; Espinosa-Aranda, 1995), Vrancea earthquakes offer the advantage of epicentral stationarity.

4. Cost-Benefit Analysis

A cost-benefit analysis should only refer to material losses and thus does not comprehensively answer the question as to whether it is desirable to become implemented. However, such an analysis can provide guidelines for the efficiency of the design.

For cost-benefit evaluation we apply the procedure outlined in Holden et al. (1989). We assume installation costs of US $ 200,000 (accelerometers: $ 20,000; computers: $ 40,000; radio communication equipment: $ 12,000; labour: $ 100,000; other costs: $ 28,000) and an amortising period of 10 years so that the hardware has to be replaced every 10 years. Those costs are fairly high estimates as some of the required equipment is already installed. The amortised annual capital costs of the system (C_s) amount to US $ 20,000. Annual operating costs (C_o) are expected as US $ 20,000 (labour: $ 10,000; communication costs: $ 2,000; others: $ 8,000). If we ignore a calculation of interests the net benefit (NB) of the system can be expressed as:

$$NB = P(eq) \cdot B_w - [C_s + C_o + P(f_a) \cdot C_{fa}] \qquad (3)$$

With the annual probability of a damaging earthquake P(eq), the estimated savings if a warning is issued B_w, the annual probability of a false alarm $P(f_a)$, and the costs of a false alarm, C_{fa}. Equation (3) contains potential annual benefits and subtracts all costs. Obviously, if NB < 0 the system is not beneficial, but if NB > 0 a net benefit can be expected. Assuming a damaging earthquake every 50 years and a false alarm every 5 years, the break-even point (NB = 0) is specified as

$$B_w = 2 \text{ Million} + 10 \cdot C_{fa} \qquad (4)$$

Thus the savings earned by a warning must exceed ten times the costs of a false alarm plus $ 2 million US. If no false alarms occur, the savings must exceed only $ 2 million US. As the damage caused in Bucharest by an $M_w > 7$ Vrancea earthquake can easily exceed that value, it means that an EWS as proposed in this paper could be cost-efficient. The estimated direct economic damage of the 1977 earthquake have been estimated as 800 million US$ by Munich Re (Forschungsgruppe Geowissenschaften, 1998) and 2 billion US$ by the World Bank. No information is currently available on how much of this amount could have been saved by an operating EWS.

5. A New Paradigm for Early Warning Systems

In recent years two novel aspects entered the early warning debate. The first relates to the thorough analysis of large-scale earthquake disasters and associated shortcomings in availability of information before, during and after a catastrophe (Comfort, 2000a; Comfort 2000b). To a large extent earthquake disasters are aggravated by the lack of information for days. It thus becomes important to develop an information infrastructure for disaster management that provides rapid information before, during and after a disaster. In terms of a temporal hierarchy this starts with literal early warning a few seconds before the disaster strikes. Then shake maps provide near real-time information on the level of ground motion within minutes. As a next step damage estimates are provided based on previously developed models of vulnerability. These projections are continuously updated by observations from the field (air borne photos, ground reports, etc.). In this frame early warning is one component of a more comprehensive system – the disaster information tool.

The second novel aspect stems from a conception of early warning that is operational rather than technical. In this view early warning systems must be made up of a number of integrated sub-systems:

- A warning sub-system, in which hazards are monitored and forecasted, at the international, national and local levels. In these, scientific information about impending hazards is produced and communicated to national authorities responsible for disaster management.
- A risk information sub-system, which can enable disaster management authorities to generate risk scenarios. These should indicate the potential impact of an impending hazard event on specific vulnerable groups and sectors of the society.
- A preparedness sub-system, in which disaster preparedness strategies are developed that indicate actions required to reduce the loss and damage expected from an impending hazard event.
- A communication sub-system, which allows the communication of timely information on impending hazard events, potential risk scenarios and preparedness strategies to vulnerable groups, so that they may take appropriate mitigation measures.

As such, an early warning system is much more than a scientific and technical instrument for forecasting hazards and issuing alerts. It should be understood as an information system designed to facilitate decision-making, in the context of national disaster management agencies, in a way that empowers vulnerable sectors and social groups to mitigate the potential losses and damages from impending hazard events. The usefulness of an early warning system should be judged, less on whether warnings are issued per se, but rather on the basis of whether the warnings facilitate appropriate and timely decision-making by those people who are most immediately at risk.

6. Conclusions

Earthquake early warning systems have the inherent weakness of short warning times. They usually amount to not more than a few seconds. Under favourable circumstances their time can be extended to half a minute (Bucharest) or even to a minute (Mexico City). Warning times can be significant if the system utilises specific tectonic circumstances of strong seismicity. They should be built in a robust way as they are expected to operate for decades in a reliable way. Real-time data communication is essential. Despite of the short warning times a number of potential applications can be specified, some of them are operated or under consideration.

EWS should be viewed as part of a general real-time information system that provides rapid information to the public and disaster relief organisations before (early warning) and after (e.g. shake maps and damage projections) the disaster strikes.

Acknowledgements

The study has been supported by the Collaborative Research Center 461 'Strong Earthquakes: A Challenge for Geosciences and Civil Engineering' of Karlsruhe University, funded by Deutsche Forschungsgemeinschaft (DFG).

References

Comfort LK (2000a) Information technology and efficiency in disaster response: The Marmara, Turkey Earthquake, 17 August 1999.
    http://www.colorado.edu/hazards/ qr/qr130/qr130.html
Comfort LK (2000b) Response operations following the Chi-Chi (Taiwan) earthquake: Mobolizing a rapidly evolving, interorganizational system. Journal of the Chinese Institute of Engineers 23:4:479-492
Espinosa Aranda JM, Jiménez A, Ibarrola G, Alcantar F, Aguilar A, Inostroza M, Maldonado S (1995) Mexico City seismic alert system. Seismological Research Letters 66:42-53

Forschungsgruppe Geowissenschaften der Münchener Rückversicherung (1998) Weltkarte der Naturgefahren, Münchner Rückversicherungsgesellschaft. 80791 München Germany

Harben PE (1991) Earthquake alert system feasibility study. Lawrence Livermore National Laboratory, Livermore, CA, UCRL-LR-109625

Heaton TH (1985) A model for a seismic computerized alert network. Science 228:987-990

Holden R, Lee R, Reichle M (1989) Technical and economic feasibility of an earthquake warning system in California. Special Publ. 101, California Dept. of Conservation, Div. Mines and Geology, Sacramento

Lee WHK, Shin TC, Teng TL (1996) Design and implementation of earthquake early warning systems in Taiwan. Proceedings 11th World Conference on Earthquake Engineering, Acapulco, Mexico

Lungu A, Coman O (1994) Experience database of Romanian facilities subjected to the last three Vrancea earthquakes. Part I: Probabilistic hazard analysis to the Vrancea earthquakes in Romania. Research Report for the International Atomic Energy Agency, Vienna, Austria, Contract No. 8223/EN

Nakamura Y (1989) Earthquake alarm system for japanese railways. Japanese Railway Engineering 28:4

Nakamura Y (1996) Real-time information system for hazard mitigation. Proceedings 11th World Conference on Earthquake Engineering. Acapulco, Mexico

Noda S Meguro K (1995) A new horizon for sophisticated real-time earthquake engineering. Journal of Natural Disaster Science 17:2:13-46

Oncescu MC Bonjer KP (1997) A note on the depth recurrence and strain release of large Vrancea earthquakes. Tectonophysics 272:291-302

Oncescu MC Marza VI, Rizescu M, Popa M (1999) The Romanian earthquake catalogue between 984-1996, In: Wenzel F, Lungu D, Novak O (eds.) Vrancea Earthquakes: Tectonics. Hazard and Risk Mitigation, Kluwer Academic Publishers, Dordrecht, Netherlands, pp 43-45

Shin TC, Tsai YB, and Wu YM (1996). Rapid response of large earthquake in Taiwan using a realtime telemetered network of digital accelerographs. Proc. 11th World Conf. Earthq. Eng., Paper No. 2137.

Teng TL, Wu L, Shin TC, Tsai YB, and Lee WHK (1997). One minute after: strong motion map, effective epicenter, and effective magnitude. Bulletin of the Seismological Society of America 87:1209-1219

Wenzel F., Oncescu MC, Baur M, Fiedrich F, Ionescu C (1999) An early warning system for Bucharest. Seismological Research Letters 70:2:161-169

Wu YM, Shin TC, Chen CC, Tsai YB, Lee WHK, and Teng TL (1997) Taiwan rapid earthquake information release system. Seismological Research Letters 68:931-943

Zschau J, Küppers, AN (eds.)(2003) Early Warning Systems for Natural Disaster Reduction, EWC '98. Springer Verlag, ISBN 3-540-67962-6, 834pp

A Combined Geophysical/Engineering Approach for the Seismic Safety of Long-Span Bridges

Andreas Fäcke[1], Lothar Stempniewski[1], Sandra M. Richwalski[2],
Stefano Parolai[2], Claus Milkereit[2], Rongjiang Wang[2], Peter Bormann[2],
and Frank Roth[2]

[1] Institut für Massivbau und Baustofftechnologie, University of Karlsruhe
[2] GeoForschungsZentrum Potsdam

Abstract

The spatial distribution of the resonance frequency of the sedimentary cover in the Cologne area was estimated by the horizontal-to-vertical spectral ratio method using ambient noise measurements. A comparison with the eigenfrequencies of selected long-span bridges crossing the Rhine river indicated an overlap in frequency and therefore potential damage in the case of an earthquake. Consequently, we conducted a dynamic vulnerability analysis using the finite-element method. In addition to accelerograms from design spectra, ground motion scenarios with sources located on the most proximate fault (Erft fault) were simulated using a numerical hybrid method that takes into account source, path, and site effects. The results indicated that due to the low frequency content of these scenarios, consistently higher responses – compared to the recommended loads in the German seismic code - were obtained at all selected bridges. Usage of both scenarios revealed specific failure mechanisms. While cable supported bridges seem to be secure, grave failure was detected for a box girder bridge.

1. Introduction

Cologne, located in the tectonically active region of the Lower Rhine Embayment, is one of the major cities in Germany. The last damaging earthquake was the 1992 Roermond earthquake with a magnitude of M_w=5.4 (Camelbeeck et al. 1994) and the epicentre located 80 km to the NW of Cologne. The damage was comparatively small and amounted to 40 million US\$. However, only 15 km west of the city is the closest fault of a potentially hazardous fault system, the Erft fault system. This fault system is believed to be capable of producing an earthquake with a magnitude of M_w=6.3 every 4900 years (Ahorner 2001). Assessing the seismic hazard as well as the vulnerability of man-made structures is one of the goals of the DFNK, which stands for Deutsches Forschungsnetz Naturkatastrophen (German Research Network for Natural Disasters). The research presented in this paper is part of this interdisciplinary project.

The city of Cologne is built on sediments of Tertiary and Quaternary age that vary up to several hundred meters in thickness. Generally, sediments play an important role for the damage potential of earthquakes causing considerable variations of the frequency-dependent amplitude amplification of seismic waves over short distances. The estimation of the spatial distribution of the resonance frequency of these sediments as well as the amplification factor linked to it are important steps in the seismic hazard assessment, since an overlap of this frequency with eigenfrequencies of man-made structures hints to an increased damage potential. Therefore, the dynamic response of the structures has to be evaluated. Generally, accelerograms obtained from design spectra or representative recordings are used. For regions, where strong motion recordings are sparse, ground motion scenarios considering source, path, and site effects, constitute an important alternative for the vulnerability analysis of man-made structures. The bridges connecting the quarters east and west of the Rhine River are a key research object, because they act as lifelines in case of a catastrophic event and have therefore to be evaluated in terms of earthquake resistance.

We will first present the results from ambient noise measurements for estimating the fundamental resonance frequency of the soil and the amplification factors. Second, we will provide information on the hybrid method we used for modelling the ground motion scenarios and show results from two scenarios. Third, we will present the eigenmodes for three bridges crossing the Rhine and compare their eigenfrequencies with that of the sediments. Finally we will evaluate their earthquake resistance in terms of the dynamic response in critical elements to design and scenario accelerograms.

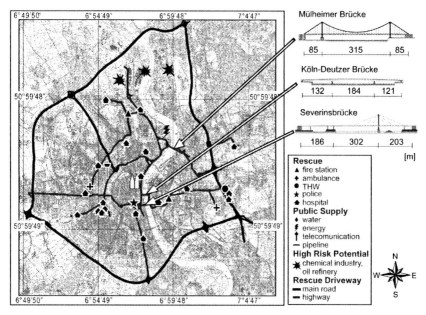

Fig. 1. Map of Cologne showing the rescue driveway net connecting important facilities for disaster management and selected bridges acting as lifelines in case of a catastrophic event

2. Assessment of Site Effects

2.1 Data Acquisition

From 12 June to 12 July 2000, noise measurements were carried out at 337 sites in the Cologne area using digital PDAS-100 TELEDYNE Geotech recording stations and Mark L-4C-3D sensors. Guralp CMG-40T sensors were used at 36 additional sites. At eight sites, noise measurements were carried out with both sensors. At each station, the signal was recorded at a sample rate of 100 Hz for at least 30 minutes. More details regarding station installation and data acquisition can be found in Parolai et al. (2001).

A second experiment was carried out from 23 April to 13 June 2001, which aimed at enlarging the number of measurement sites and at recording local and regional earthquakes. A network of 44 stations was deployed. At 40 sites PDAS-100 TELEDYNE Geotech stations equipped with Mark L-4C-3D sensors were installed. Four sites were equipped with Guralp CMG-3ESPD recording stations using Guralp CMG-3T sensors. At each PDAS-site signals were recorded at a sample rate of 100 Hz whereas a sample rate of 50 Hz was used at Guralp-sites. All stations recorded continuously. Consequently, a large amount of ambient noise data was recorded. More details regarding station installation and data acquisition can be found in Parolai et al. (2002b).

2.2 Data Analysis

The H/V ratio technique (Nakamura 1989) was used to estimate the fundamental resonance frequency of the sedimentary cover. This method requires computing the spectral ratio between horizontal and vertical components (H/V) of seismic ambient noise. The noise recordings were divided into 60 s windows, avoiding badly acquired signals. Therefore, the number of windows selected for each site ranged between 5 and 26. Note, that more than 10 windows were obtained for 95% of the sites. In a second step, each time series was tapered with a 5% cosine function and the Fast Fourier Transform (FFT) was calculated for each component. Spectra were corrected for instrumental response and then smoothed using a Hanning window with constant relative bandwidth of 28%. Third, the spectra of the NS and EW components were merged to obtain a horizontal component spectrum by means of computing their root-mean square average. Then the spectral ratios between the merged horizontal and vertical components were calculated. Finally, the H/V ratio site responses were estimated computing the arithmetical mean of all the spectral ratios calculated for the respective site.

2.3 Results

Using the method described above we estimated the fundamental resonance frequency of the soft soil cover at over 400 sites. This allowed creating a map of the fundamental frequencies in the Cologne area. In order to better determine the limits of the main spatial variations of the fundamental resonance frequency we used an adjustable-tension continuous curvature surface gridding algorithm (Wessel and Smith 1991). It allows obtaining a continuous field of the fundamental frequency (Fig. 2). In general, the spatial variation of the fundamental resonance frequency depicted in Fig. 2 is consistent with the map previously published by Parolai et al. (2001), but better constrained due to the larger number of measurement sites. Very low values of the fundamental resonance frequency (around 0.2 Hz) always appear west of the Erft fault system. In this area, the Devonian basement can be deeper than 1000 m (Von Kamp 1986; Parolai et al. 2002a).

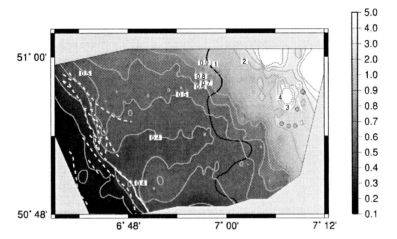

Fig. 2. Map of the fundamental resonance frequency (Hz) of the soil cover in the studied area. Grey circles indicate stations with flat H/V ratio site responses. Note that the fundamental resonance frequency field is well constrained only in areas with a high density of neighbouring points (Fig. 3). Fault positions are taken from Von Kamp (1986).

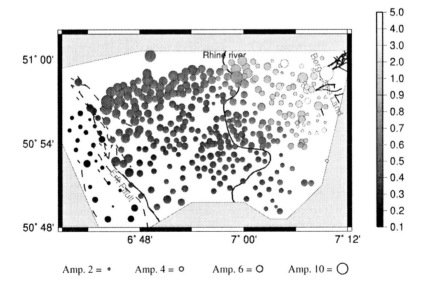

Amp. 2 = •   Amp. 4 = o   Amp. 6 = O   Amp. 10 = ◯

Fig. 3. Map showing the fundamental resonance frequencies (Hz) measured at all the analysed sites. Triangles indicate stations with flat H/V ratio site responses. The radius of the circles is proportional to the observed amplification factors. Fault positions are taken from Von Kamp (1986).

It is worth noting the agreement between the spatial variation of the fundamental resonance frequency and fault positions. Much higher frequencies (between 1 and 5 Hz) are observed NE of the Rhine river towards the outcropping Devonian basement in the Bergisches Land. In the urban area of Cologne, the fundamental resonance frequencies range between 0.4 and 1 Hz. These results are in agreement with previous studies (Ibs-von Seht and Wohlenberg 1999; Parolai et. al 2001, 2002a), which showed that H/V spectral ratio of ambient noise could be used to assess variations in the thickness of sedimentary cover.

The amplification factors obtained by the H/V ratio of ambient noise in the studied area are shown in Fig. 3. Although the representativeness of the H/V amplitudes for assessing S-wave amplification is still debated, the picture shows that amplification values may range at least between 2 and 12. On average, the H/V ratio tends to underestimate earthquake site amplification compared to techniques, which use earthquake recordings and therefore directly estimate the S-wave amplification (e.g Bard et al. 1997).

It is also worth noting that inside the basin the highest values of amplifications generally occur in the northern side of the studied area in the range 0.4-0.7 Hz. Moreover, where the resonance frequencies are very low ($\otimes$0.2 Hz) the amplifications are small too ($\otimes$ 6). This could be explained by considering that in this area of thick sedimentary cover the main impedance contrast (contact between sediments and bedrock) is reduced. Additionally, larger seismic wave attenuation may occur in areas of thick sedimentary cover.

2.4 Ground Motion Scenarios

Laterally varying structures like the Lower Rhine Embayment, which is characterised by a horst-graben structure with fault systems trending SE-NW, may have a significant effect on the propagation of seismic waves. The computation of ground motion scenarios should use at least 2D methods to allow for rather realistic modelling, since source, path, as well as site effects can be included.

We use a hybrid approach, combining an improved Thomson-Haskell propagator matrix method (Wang 1999) for the computation in a 1D background medium with a 2D viscoelastic finite-difference (FD) program of the order O(2,2) (Zahradník and Moczo 1996). The advantage of using such a hybrid approach is that the time-consuming FD-modelling is restricted to the area where the lateral variation of the physical parameters is important. Note, that in the present version of the software the source is implemented as a point source. We selected a profile simplified after a geological section (Budny 1984), which crosses the Lower Rhine Embayment roughly in East-West direction and goes through Cologne (Fig. 4, top). The velocity model of the uppermost 2 km of the crust is shown in Fig. 4, bottom.

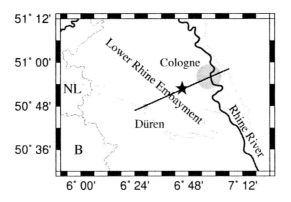

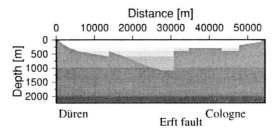

Fig. 4. Top: Location of the modified geological profile (solid straight line) and the scenario earthquake (star) on the Erft fault that runs roughly SE-NW. Bottom: The dotted lines represent the approximate outcropping of the Devonian bedrock (Hollnack, pers. comm.). The velocity model used for the scenarios. Light grey colours refer to the velocity structure of the sediments, dark grey ones to that of the background model.

The layer parameters were computed from velocity depth functions presented in Budny (1984): $vp(z)=1324\ (1+z)^{0.08}$ and $vs(z)=166\ (1+z)^{0.271}$, where z is the depth in m. The P-wave velocity background model was taken from Ahorner and Pelzing (1985) and the ratio between P- and S-wave velocities is $vs=vp/\sqrt{3}$. Q-values inside the sediments increase with depth from 40 to 60 and are 140 in the background model. Using a grid spacing of 19.5 m in the FD-modelling allows for including frequencies up to 4 Hz. The number of grid cells was 2797x116. Surface topography was not included, but is also not significant in the region under study.

The maximum magnitudes expected at the Erft fault system for different return periods are still subject to on-going research. Therefore, we base our choice of parameters for the scenarios on that of historical events on other faults in the Lower Rhine Embayment (e.g. Roermond 1992; Liège 1983). The parameters used for two different scenarios (normal faulting/ strike-slip) are shown in Table 1. We used the same moment magnitude M_0, which is equivalent to an M_w of 4.7 and therefore rather small, and depth for both scenarios to show the influence of the source mechanism. Fig. 5 displays the ground velocity on the NS (transversal component) for strike-slip and normal faulting scenarios at 91 hypothetical stations placed equi-distantly along the profile shown in Fig. 4. We have chosen global ampli-tude scaling for the two figures to make them comparable. The normal faulting scenario produces generally lower amplitudes on the NS compo-nent than the strike-slip scenario but higher ones on the EW and vertical components. The long time duration of the signal is mainly due to waves being reflected back at dipping interfaces. Moreover, a sonogram analysis (Richwalski et al. 2002) demonstrated that this part of the signal has a very narrow frequency bandwidth. In case of an earthquake elongated man-made structures like bridges might suffer heavy damage due to fatigue of material if their eigenfrequencies lie within this frequency band. The fo-cusing effect (caused by constructively interfering waves) visible in Fig. 5 before kilometre 50 can be traced back to a re-thickening of the sediments east of the Erft fault. Such a structure is confirmed to be present in the southern part of the city of Cologne according to the map of the sedimen-tary thickness derived by Parolai et al. (2002a).

The importance of at least 2D modelling for the present structure - com-pared to 1D modelling - is demonstrated with Fig. 6. It shows the NS com-ponent of the ground velocity obtained from 1D modelling for the normal faulting scenario. We mimicked the 2D structure in Fig.4 (bottom) by re-peating the modelling for each receiver location with the corresponding 1D profile. Evidently, the signal duration is much shorter, if the waves re-flected back from dipping interfaces are missing. Since we have only lim-ited information on the geological structure we need to compare our mod-elling results with measured data to gain confidence in our results. Lacking earthquake recordings, we computed the spectral ratio between the com-bined horizontal components and the vertical ones from the 2D normal faulting scenario to compare them with the resonance frequencies obtained from ambient noise measurements.

Table 1. Source parameters for the two earthquake scenarios at the Erft fault

scenario	depth	M_0 [Nm]	strike	dip	rake
strike-slip	6 km	$1.6\ 10^{16}$	347°	60°	12°
normal faulting	6 km	$1.6\ 10^{16}$	130°	70°	-100°

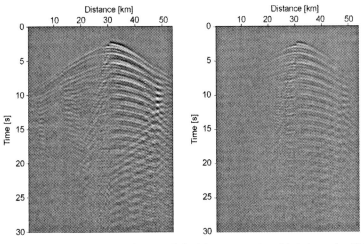

Fig. 5. Strike-slip (left) and normal faulting scenarios (right) on the Erft fault, NS component of ground velocity. Amplitudes are comparable.

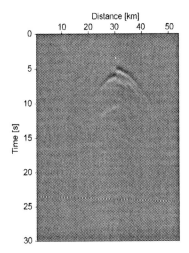

Fig. 6. Strike-slip scenario on the Erft fault, Normal faulting scenario on the Erft fault, 1D realisation of the normal faulting scenario on the Erft fault, NS component of ground velocity. Amplitudes are comparable to Fig.5.

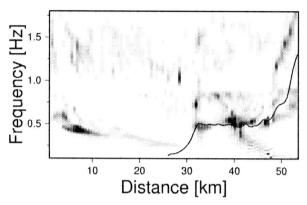

Fig. 7. Spectral ratios between the combined horizontal and vertical components of the seismograms for the normal faulting scenario. The black line shows the estimated values.

Fig. 7 shows the spectral ratios using the complete seismograms at each receiver position. Predominant frequencies appear in darker grey colour than less significant ones. Evidently, the thicker the sediments are the lower the fundamental frequency becomes. Also higher modes are visible. For comparison, the estimated fundamental frequencies (Parolai et al., 2001) are shown as well and there is generally a good agreement. Close to the Erft fault (between kilometre 25 and 30) our model exhibits higher frequencies than the measured data. This is an indication that we have to increase the sediment thickness west of the Erft fault in future models. In addition, we expect further improvements by updating our velocity-depth-model inside the sediments, since Parolai et al. (2001a) recalibrated the velocity depth function of Budny (1984) for the region of Cologne.

3. Vulnerability Assessment of Bridges

3.1 Modelling

A modal analysis including resonance frequencies, mode shapes as well as dynamic simulations for earthquake scenarios was performed for three Rhine bridges using the finite element (FE) method. All analyses were carried out with linear elastic material behaviour while geometrical non-linear effects were considered for the cable suspension bridge due to possibly large deformations. The FE analysis program ABAQUS version 5.8 was used for all calculations. The bridges were modelled in 3D space according to descriptions in the literature (Schüßler and Pelikan 1951; Knop 1979;

Heß 1960) and structural plans from the bridge authority of Cologne 'Amt für Brücken und Stadtbahnbau'.

The *Köln-Deutzer Bridge* is a 473 m long box girder bridge with two parallel superstructures one made of steel and the other of lightweight pre-stressed concrete. The bridge is discretised with 337 beam elements (Fig. 8). The main focus is set on the lightweight pre-stressed concrete superstructure, which is modelled with beam elements. The steel superstructure is simply considered by lumped masses connected to the piers and abutments. Because of short and stiff piers, soil structure interactions are considered according to a model of Wolf (1988).

The *Mülheimer Bridge* is a cable suspension bridge with a steel superstructure and a length of 485 m. It is discretised with 1741 elements. Pylons and superstructure are modelled with beams, the deck slab with membranes, and vertical hangers with truss elements. To increase numerical stability the main cable is represented by beam elements with small bending stiffness instead of truss elements. The calculations were performed geometrical non-linear considering initial conditions like stresses and deformations from gravity loads. In this way the variation of the structural stiffness due to large deformations is incorporated.

The *Severinsbrücke* is a cable-stayed bridge made of steel with a total length of 691 m. It is modelled with 1695 elements with beams for the pylons and superstructure, membranes for the deck slab, and trusses for the cables. Each cable is represented by one element, neglecting unessential cable vibrations. A reduction in stiffness because of the cable sag is considered according to Kim et al. (2001).

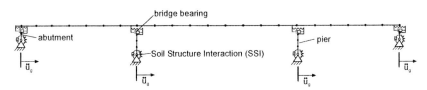

Fig. 8. Finite element model of Köln-Deutzer Bridge

3.2 Resonance Frequencies

A first estimate of the seismic damage potential of man-made structures can be made by comparing their relevant natural frequencies with the expected predominant frequency content of a seismic input. Due to amplification effects the deformations and forces within the structure highly increase. According to Fig. 2 the fundamental resonance frequency of the soil cover in the urban area of Cologne is estimated to lie between 0.4 and 1 Hz. This low frequency range is critical for high buildings like skyscrapers and towers and for long-span bridges. For this reason, the natural frequencies of vibration and the mode shapes were calculated for the selected bridges. The eigenvalue problem was solved using the subspace iteration technique implemented in ABAQUS.

Table 2. Resonance frequencies of bridges and of soil at corresponding locations

Eigenmode	Köln-Deutzer Bridge		Mülheimer Bridge		Severinsbrücke	
	f [Hz]	shape	f [Hz]	shape	f [Hz]	shape
1	0.40	vert.	0.12	vert.	0.37	vert.
2	0.75	vert.	0.28	vert.	0.70	lateral
3	0.96	lateral	0.34	vert.	0.72	vert.
4	1.08	vert.	0.50	vert.	0.84	vert.
5	1.12	long.	0.61	lateral	0.91	torsional
soil	0.55		0.75		0.50	

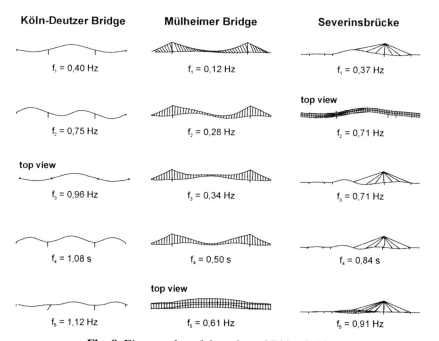

Fig. 9. Eigenmodes of the selected Rhine bridges

Having a small influence on frequencies and mode shapes, the structural damping was neglected during the eigenvalue extraction. The first five eigenmodes for each bridge are shown in Fig. 9 and are listed in Table 2. The 1st, 2nd, and 4th eigenmodes of the *Köln-Deutzer Bridge* represent a vertical swinging of the superstructure. Because the bridge is generally designed for vertical loads, these mode shapes have a subsidiary importance for seismic excitation and are listed for completeness only. The 3rd mode is a lateral bending of the superstructure, which produces high lateral forces in the bridge bearings. The bridge bearings are the connection between either the superstructure and the piers or the superstructure and the abutments. The 5th eigenmode is predominantly a translational displacement of the superstructure in longitudinal direction due to an inclination of the pier. It also produces high forces in the bridge bearings, but in longitudinal direction. According to Fig. 2 the resonance frequency of the sediments at the bridge is 0.55 Hz. The relevant natural frequencies of the bridge are 0.96 Hz for the 3rd and 1.12 Hz for the 5th mode. Though resonance frequencies of the bridge and the sediments do not overlap, they are close, which results in a moderate amplification.

The 1st eigenmode of the *Mülheimer Bridge* is a pendulum-like swinging of the superstructure in longitudinal direction. It produces negligible forces within the structure. The 2nd to 4th eigenmode are vertical bendings of the superstructure again with subordinate importance to seismic excitations. The 5th mode represents a lateral movement of the main cable and superstructure especially in the middle field. This eigenmode is important for a lateral seismic excitation. At the *Mülheimer Bridge* the soil resonance frequency is 0.75 Hz while the relevant 5th mode of the structure amounts to 0.61 Hz. Consequently, an intense amplification in lateral direction is to be expected.

The 1st, 3rd, and 4th eigenmode of the *Severinsbrücke* are characterized predominantly by a vertical bending of the superstructure. In contrast to the other bridges, these vertical eigenmodes are important for a cable-stayed bridge. The 2nd mode is relevant for a lateral ground motion. The 5th is a torsional mode of the superstructure with minor importance for seismic input. At the corresponding location the site fundamental frequency is 0.5 Hz while the relevant frequencies of the bridge are 0.37 to 0.84 Hz. High amplifications in both horizontal directions are to be expected.

Except for the longitudinal direction of the *Mülheimer Bridge* the soil and structural resonance frequencies are quite close for all bridges. Therefore, high amplifications can be expected for these long-span bridges. However, high deformations and forces in the structure due to amplification do not necessarily cause damage to the structure. A well-designed structure with high strength and/or deformation capacity is able to withstand seismic excitation even if the resonance frequencies of the structure and the sediments overlap. For this reason, the dynamic response of the structure has to be calculated for realistic scenarios.

3.3 Dynamic Simulations

In order to get a precise response of buildings to specific accelerograms the time integration method has to be used. With this method effects like plastic deformations, isolation devices, stiffness, and strength degradation of material from repeated cycling is possible. One accelerogram was generated with Simqke I consistent to the design spectrum of the new German code E-DIN 4149, which actually is in design stage. This design spectrum considers site effects, though constant for the entire city. The duration of strong motion is set to 5 sec according to Hosser (1987). An envelope function from Klein (1985) with constant plateau during strong motion phase is used leading to a total duration of 15 sec.

Table 3. Maximum Response of the bridges to selected scenarios

maximum response	Köln-Deutzer Bridge	Mülheimer Bridge	Severinsbrücke
strike-slip	3.40	0.58	2.75
E-DIN 4149	1.80	0.40	2.60
bearing capacity	0.5	-	3.25

The peak ground acceleration is scaled to 0.52 m/s² conformable to E-DIN 4149 for bridges with high importance in seismic hazard zone 1. In addition, the strike-slip scenario simulated with the hybrid method was used. We took one trace of the NS-component at the station nearest to the bridges without any scaling. It has a total duration of 40 sec, a strong motion phase of 23.4 sec and a peak ground acceleration of 0.60 m/s².

The strong motion phase is defined according to Kennedy with 70 % of the total energy spaced between 5 and 75 %, which correlates best with structural damage (Hosser 1987).

The time history analysis was performed using the implicit direct integration method implemented in ABAQUS. A time step of 0.01 s was used, which allows including resonance frequencies up to 100 Hz. The accelerograms, time histories of the response (Fig. 10) and the results of the dynamic simulations (Table 3) at one representative part at each bridge, which we rated most critical, are shown. At the *Severinsbrücke* the acting force in the connection between pylon and superstructure (fixed bearing) for an excitation in longitudinal direction can be seen. For both scenarios

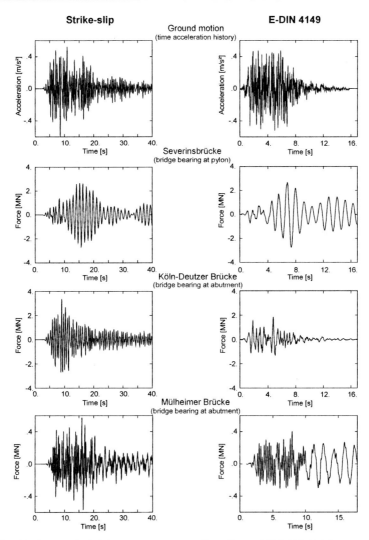

Fig. 10. Dynamic bridge response to strike-slip and E-DIN 4149 scenario

the maximum force in the connection is about 2.7 MN. With an ultimate load capacity of 3.25 MN there is no damage to expect, however there is only little scope left. The *Köln-Deutzer Bridge* was excited in transversal direction and the force in the connection between superstructure and abutment (thrust bearing) are measured. The maximum force from the simulation with the strike slip scenario is 3.4 MN and with the E-DIN scenario it is 1.8 MN. With an ultimate load of 0.5 MN the connection will break for both scenarios. The *Mülheimer Bridge* was excited in lateral direction and the force in the anchorage between abutment and superstructure is meas-ured. The strike-slip scenario again yields with 0.58 MN a higher response than the E-DIN scenario with 0.40 MN. An ultimate load could not yet de-termined but the response seems rather small.

However in this special case the damage prognosis is the same the strike slip-fault scenario generally yields greater response than the E-DIN sce-nario. The difference is greater than just the difference in peak ground ac-celeration, which is a commonly employed indicator for the severity of a ground motion.

4. Conclusions

We assessed the vulnerability of three long-span bridges crossing the Rhine River at Cologne by taking into account source, path as well as site effects. The bridges (Mühlheimer Bridge, Severinsbrücke, and Köln-Deutzer Bridge) will act as lifelines in case of a catastrophic event, and their resistance to ground shaking due to earthquakes have to be evaluated.

Using the H/V-ratio technique, the spatial distribution of the fundamen-tal resonance frequency of the soil was estimated. Values of the fundamen-tal resonance frequency of 0.4 to 1 Hz in the urban area indicate a high vulnerability potential for the selected long-span bridges, which have rele-vant eigenmodes with eigenfrequencies close to the estimated ones. We conducted a dynamic simulation to investigate the influence of the site ef-fect on specific damage patterns at the bridges. In addition to a scenario from the German code E-DIN 4149 we used one modelled by a 2D hybrid method, which takes into account effects of the source, path and site. Al-though both scenarios yield in this specific case the same damage estima-tion, namely break of bridge bearings at the box girder bridge, the strike-slip scenario with a lower frequency content than the E-DIN scenario led to notable higher responses for all bridges.

The results show that local site conditions play an important role in the vulnerability assessment, particularly for special structures like long-span bridges. Therefore a mere analysis according to the seismic code seems not to be sufficient for this kind of structures. Note, that the ground motion scenarios modelled so far did not simulate the maximum possible event that might be expected at the Erft fault. Therefore, further investigations have to be done in the field of possible scenarios concerning the city of Cologne. Also the influence of the spatially varying ground motion input along the bridges, which is rated to be generally harmful, has to be investi-gated.

Acknowledgements

Mr. R. Thon and Mr. K. H. Hasse from the Cologne building authority 'Amt für Brücken und Stadtbahnbau' kindly made structural plans available to us. This study was carried out with the support of the BMBF under contracts No. 01SF9969/5 and 01SF9973/0. The Geophysical Instrument Pool Potsdam (GFZ) provided instruments. Figures were generated using Seismic Unix (SU) and GMT (Wessel and Smith, 1991).

References

Ahorner L (2001) Abschätzung der statistischen Wiederkehrperiode von starken Erdbeben im Gebiet von Köln auf Grund von geologisch-tektonischen Beobachtungen an aktiven Störungen. DGG Mitteilung 2: 2-10

Ahorner L, Pelzing R (1985) The source characteristics of the Liège earthquake of November 8, 1983, from digital recordings in West Germany. in P Melchior (ed), Seismic Activity in Western Europe, D Reidel Publishing Company: 263-289

Bard P-Y, Duval A-M, Lebrun B, Lachet C, Riepl J, Hatzfeld D. (1997) Reliability of the H/V technique for site effects measurement: An experimental assessment. Proc. 8th Int. Conf. Soil Dyn. Earthq. Engrg., Istanbul, Turkey

Bathe KJ (1996) Finite Element Procedures. Prentice Hall, New Jersey

Budny M (1984) Seismische Bestimmug der bodendynamischen Kennwerte von oberflächennahen Schichten im Erdbebengebiet der Niederrheinischen Bucht und ihre ingenieurseismologische Anwendung. Ph.D. thesis, Geologisches Institut der Universität zu Köln

Camelbeeck T, van Eck T, Pelzing R, Ahorner L, Loohuis J, Haak HW, Hoang-Trong P, Hollnack D (1994) The 1992 Roermond earthquake, the Netherlands, and its aftershocks. Geologie en Mijnbouw 73: 181-197

E-DIN 4149 (2000) Auslegung von Hochbauten gegen Erdbeben (Entwurf). Deutsches Institut für Normung, NaBau, Berlin

Eurocode 8 Design provisions for earthquake resistance of structures. ENV 1998-1-1 (1994), ENV 1998-1-2 (1994), ENV 1998-1-3 (1995), European Committee for Standardization, Brüssel

Heß H (1960) Die Severinsbrücke Köln – Entwurf und Fertigung der Strombrücke. Der Stahlbau 29: 225-261

Hosser D (1987) Realistische seismische Lastannahmen für Bauwerke – Ergebnisse einer interdisziplinären Forschungsarbeit. Bauingenieur 62: 567-574

Klein HH (1985) Kenngrößen zur Beschreibung der Erdbebeneinwirkung - Mitteilungen aus dem Institut für Massivbau der TH Darmstadt. Ernst&Sohn, Berlin

Knop D (1979) Die Verbreiterung der Rheinbrücke Köln-Deutz. Beton-Information 4: 14-35

Ibs-von Seht M, Wohlenberg J (1999) Microtremor measurements used to map thickness of soft sediments. Bulletin Seismological Society of America 89: 250-259

Nakamura Y (1989) A method for dynamic characteristics estimations of subsurface using microtremors on the ground surface. Quarterly Report RTRI, Japan 30: 25-33

Parolai S, Bormann P, Milkereit C (2001) Assessment of the natural   frequency of the sedimentary cover in the Cologne area (Germany) using noise measurements. Journal of Earthquake Engineering 5: 541-564

Parolai S, Bormann P, Milkereit C (2002a) New relationships between Vs, thickness of the sediments and resonance frequency calculated by means of H/V ratio of seismic noise for the Cologne area (Germany). Bulletin Seismological Society of America, in press

Parolai S, Richwalski SM, Milkereit C, Bormann P (2002b). Assessment of the stability of H/V spectral ratios and comparison with earthquake data in the Cologne area (Germany). submitted to Tectonophysics

Richwalski SM, Parolai S, Wang R, Roth F (2002) Shaking duration and resonance frequencies from 2D hybrid modelling for the region of Cologne (Germany), 27th General Assembly of the European Geophysical Society, paper SE079

Schüßler K, Pelikan W (1951) Die neue Rheinbrücke Köln-Mülheim. Der Stahlbau 20: 141-150

Shinozuka M, Saxena V, Deodatis G (2000): Effect of spatial variation of ground motion on highway structures. Technical Report MCEER-00-0013 Princeton University, New Jersey

Von Kamp H [1986] Geologische Karte von Nordrhein-Westfalen 1:100000. Geologisches Landesamt Nordrhein-Westfalen

Wang R (1999) A simple orthonormalization method for stable and efficient computation of Green's functions. Bulletin Seismological Society of America 89: 733-741

Wessel P, Smith WHF (1991) Free software helps map and display data. EOS. Trans. AGU 72 (41): 441, 445-446

Wolf JP (1988) Soil-structure-interaction analysis in time domain. Prentice Hall, New Jersey

Zahradník J, Moczo P (1996) Hybrid seismic modeling based on discrete-wave number and finite-difference methods. PAGEOPH 148: 21-38

Finite-Difference Simulations of the 1927 Jericho Earthquake

Ellen Gottschämmer[1], Friedemann Wenzel[1], Hillel Wust-Bloch[2], and Zvi Ben-Avraham[2]

[1] Geophysical Institute, Karlsruhe University, Germany
[2] Dept. of Geophysics and Planetary Sciences, Tel Aviv University, Israel

Abstract

Four possible scenarios of the 1927 Jericho earthquake are tested by simulating 75 seconds of 1.5 Hz-wave propagation in a 3D model of the Dead Sea Basin (DSB) substructure. The scenarios examine the effects of various source and rupture parameters, since the original parameters could not be constrained by the sparse data gathered in 1927. The simulations are carried out using a fourth-order staggered-grid finite-difference (FD) method. Peak ground velocities and spectral accelerations at 0.5 Hz, 1 Hz, and 1.5 Hz) are determined from the time-histories. Finally those are compared to an intensity map, which shows a considerable heterogeneous distribution of large intensities. The purpose of this study is (a) to find possible explanations for the heterogeneous intensity distribution and (b) to determine the best fitting source and rupture parameters for this event. We find the best overall agreement to correspond to scenario 2, a unilateral rupture on a vertical strike-slip fault with a fault plane of $12 \cdot 12$ km^2. The rupture starts at a depth of 7 km at the southern end of the fault and propagates with a velocity that amounts to 90% of the local shear wave speed.

1. Introduction

In the mid-Cenozoic, the Arabian plate broke away from the African plate, forming the Dead Sea Rift (DSR) transform at the boundary between the

Arabian and an appendage of the African plate, the Sinai sub-plate. The transform fault system extends over 1000 km, from the divergent plate boundary along the Red Sea in the south to the East Anatolian Transform in the north (Fig. 1). The Dead Sea Basin, being one of the largest pull-apart basins in the world, formed along the left-lateral DSR transform

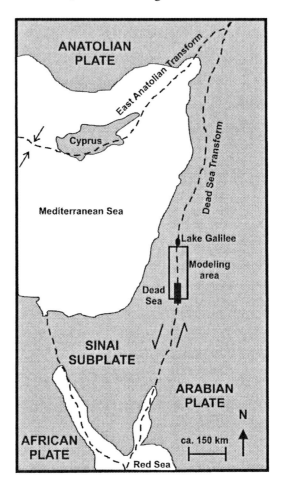

Fig. 1. Map of the Dead Sea Transform and the Dead Sea Basin (after Ben-Menahem 1991). Plate boundaries are indicated as dashed lines. The location of the Dead Sea and the Lake Galilee are sketched. The approximate location of the modeling area is indicated as a black rectangle.

fault. Subsidence of the basin occurred in large parts about 15 Ma or earlier with the site of fastest subsidence having shifted northward. By the end of the Miocene about half of the present length was reached (Garfunkel and Ben-Avraham 1996). Note that the center of the Dead Sea Basin is not located directly under the Dead Sea but extends further to the south (Ben-Avraham, 1997).

The DSR transform has been known as a source region for earthquakes with magnitudes reaching 7.5. Since the region has been settled for several centuries, first reports of earthquakes date back to more than 4000 years. Several catalogues are available for major events (Ambraseys et al. 1994; Amiran et al. 1994; Ben-Menahem 1979, 1991; Sieberg 1932). The largest magnitude events are compiled in Table 1.

Table 1. Large historic earthquakes connected to the Dead Sea Rift transform

Year	M_L (est.)	Comments
2100 BC	6.8	Upheaval in southern Dead Sea, destruction of Jericho.
759 BC	7.3	King Solomon's temple in Jerusalem damaged.
746 AD	7.3	Destruction of the Omaya Hesham palace near Jericho.
1170 AD	7.5	King Herod's obelisk at Ceasaria fell down.
1202 AD	7.5	Great damage and many victims. Felt in a distance of 1200 km.
1546 AD	7.0	Seiches in the Dead Sea. Water of river Jordan cut for 2 days.
1759 AD	7.4	Destruction of Baalbek. Parts of Damascus destroyed.
1837 AD	6.7	Destruction of Safed and Tiberias, 3000 victims.
1927 AD	6.2	Epicenter in Jericho. First event instrumentally recorded.

M_L *(est.)* estimated local magnitude

The most recent of those earthquakes, a $M_L = 6.2$ earthquake, occurred on 11 July 1927 (Shapira et al. 1993) near the town of Jericho where it destroyed several buildings and caused ground fissures (Nur and Ron 1996, Sieberg 1932). In many cities and villages in Judea, Samaria and Galilee the earthquake had even worse effects: More than 340 people were killed and almost 1000 were injured due to the collapse of buildings. The flow of the river Jordan stopped for 22 hours because of landslide failures. Shaking has been reported from regions 900 km away from the epicenter (Sieberg, 1932). The event could be recorded at several seismic stations then available in Europe, South-Africa, North-America, and even Australia, thus extending the microseismic coverage to more than 14000 km (Sieberg 1932).

Fig. 2 shows the distribution of the macroseismic intensity for the 1927 event. The intensity pattern shows a distinct extension to the north: The area where the largest damage is found (intensity = X) is situated approximately 50 km north of the epicenter on the eastern side of the Ephraim mountains and is embedded in a kidney-shaped elongated zone of intensity > VIII. Even further to the north, in a distance of approximately 100 km from the epicenter, two patches with intensity level VIII-IX are situated. The deviation from a more radial pattern of intensity distribution has been reported for several other events (Sieberg 1932; Wust-Bloch and Wachs 2000) and recently it was suggested that the elongated structure of the Dead Sea Basin acts like a wave-guide (Wust-Bloch 2002). Beyond the fact that the heterogeneous distribution of intensities might have also been enhanced by irregular clusters of population and inhomogeneous site effects, the general pattern shows a higher level of intensities to the north of the epicenter.

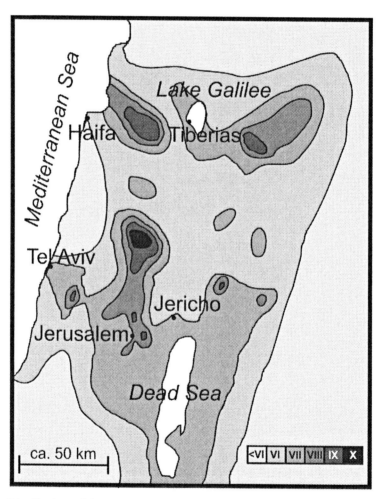

Fig. 2. Distribution of the macroseismic intensity for the 1927 Jericho earthquake (after Sieberg 1932). The maximum intensity amounts to X and is found approximately 50 km north of the epicenter. Two further patches with large intensities (IX) are found to the East and to the West of Lake Galilee

The aim of this study is to understand the heterogeneous distribution of intensities reported by Sieberg, and to quantify the effect by 3D FD modeling of ground-motion. Heterogeneous distribution of ground motion was successfully computed by 3D modeling of wave-propagation in deep sediment-filled basins e.g. for the Tokyo Metropolitan Area (Sato et al. 1999), the Los Angeles Basin (Olsen 2000), or the Upper Rhine Graben (Gottschämmer et al. 2000) and could be explained by significant focusing effects and scattering of seismic waves. A comparison between synthetic ground motion for a DSB model and a uniform half-space model showed that the geological subsurface basin structure is responsible for a major part of ground motion amplification north of the DSB (Gottschämmer et al. 2002). More precisely, 1 Hz-spectral accelerations were amplified by a factor of more than 5. While ground motion amplification could be well explained, the distribution of large ground motion amplitudes showed considerable differences in comparison to the intensity map (Fig. 2). This pattern is crucially dependent on the rupture history of the earthquake, and is examined in the present study. Since hypocentral location and rupture parameters are not known in detail for the 1927 Jericho earthquake, we compute wave-propagation for four rupture scenarios and compare the resulting distribution of peak ground motions to the distribution of intensities revealed by Sieberg.

2. Earthquake Scenarios

We choose four scenarios to compute ground-motion in the DSB area, all being implemented on a 90° dipping fault using a strike-slip source mechanism. Several source and rupture parameters, such as the hypocentral coordinates, the depth and size of the rupture area, and the rupture velocity have not been determined for the Jericho earthquake. In our simulations, we vary these parameters, in order to find a rupture model that can qualitatively explain the distribution of large intensities north of the epicenter of the 1927 Jericho earthquake. The size of the rupture area A was computed using empirical relationships for strike-slip events (Wells and Coppersmith 1994). A square with length 12 km is used for scenario 1. The rupture extends up to $z_D = 200$ m below the free surface. It propagates unilateral from the hypocenter, which we fixed at the southern end of the fault, at a depth d_H of 7 km at the edge of the basin. This point was chosen because here the fault plane crosses the transition from bedrock to the sediments. The rupture velocity v_R is a function of the local shear-wave speed and amounts to 70% v_S in accordance with observations for shallow earthquakes (Geller 1976). A different rupture velocity is used for scenario 2 where v_R amounts to 90% v_S, respectively, taking into account that recently larger rupture velocities have been observed (Fukuyama and Olsen, 2002). All other parameters are kept constant. A bilateral rupture is simulated in scenario 3. The rupture initiates at a depth of 7 km at the center of the fault plane and propagates with 90% v_S simultaneously towards north and south. All other rupture parameters correspond to those of scenario 1. A slightly larger, rectangular rupture plane with a size of 10 km by 15 km is used for scenario 4. In this simulation, the rupture plane extends from a depth of 10 km to a depth of 20 km. The rupture starts at the southern end of the fault at a depth of 15 km and propagates with 90% of the local shear wave speed, which is constant at 2.86 km/s. The rupture parameters are compiled in Table 2.

Table 2. Earthquake rupture parameters for scenario 1-4

Scenario	A (km^2)	z_D (km)	d_H (km)	v_R	type of rupture
1	12·12	0.2	7	70% v_S	UL
2	12·12	0.2	7	90% v_S	UL
3	12·12	0.2	7	90% v_S	BL
4	10·15	10.0	15	90% v_S	UL

A size of rupture area, z_D depth to top of rupture plane, d_H hypocentral depth, v_R rupture velocity, v_S shear wave velocity, UL unilateral, BL bilateral.

3. Elastic and Computational Modeling Parameters

For our simulations, a simplified 3D elastic model is used, as proposed for the DSB area (Al-Zoubi et al. 2002; Garfunkel and Ben-Avraham 1996; Ginzburg and Ben-Avraham 1997) (Fig. 3). It consists of a basin, $15 \cdot 100$ km^2 in size, with three layers that are embedded in crystalline bedrock (compressional wave velocity $v_p = 5.5$ km/s, shear wave velocity $v_s = 3.2$ km/s, density $\rho = 2.7$ g/cm^3). The upper layer with $v_p = 4$ km/s, $v_s = 2.3$ km/s, and $\rho = 2.1$ g/cm^3 corresponds to shales and clastics and extends to a maximum depth of 5 km. The two underlying layers extend in the center of the basin to a depth of 9 km and are composed of salt and quartzrose sandstones ($v_p = 4.5\text{-}5$ km/s, $v_s = 2.6\text{-}2.9$ km/s and $\rho = 2.4\text{-}2.6$ g/cm^3). Small-scale features in the geometry of the basin have been deliberately ignored in order to keep the model as simple as possible. Our simulations have been restricted to frequencies below 1.5 Hz. The relatively low frequencies correspond to a minimum wavelength for shear waves of 1.5 to 2.0 km. Model features below this scale are thus considered as unimportant. Changes in the model of this scale size would not affect the results significantly. For the same reason no mountain topography has been included in the model. The simulations comprise no material attenuation since the effect is not of first order at the frequencies and distances considered.

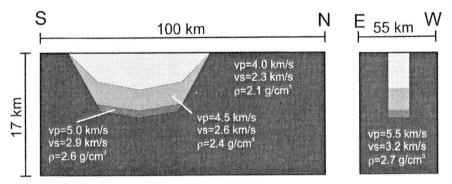

Fig. 3. Two cross-sections of the model of elastic parameters used for the simulation of the Jericho earthquake. Note that the vertical axis is exaggerated by a factor of 4.

To solve the 3D elastic equations of motion a staggered-grid FD scheme is used (Olsen 1994). The accuracy is fourth-order in space and second-order in time. The numerical implementation of the free-surface boundary condition is described in (Gottschämmer and Olsen 2001). In order to implement the source to the FD grid, $\Delta t/\Delta x^3 \cdot dM_{ij}/dt$ is added to the ijth component of the stress tensor σ_{ij}, where M_{ij} is the ijth component of the moment rate tensor, and Δt and Δx denote the temporal and spatial discretization, respectively. All four scenarios use a triangular-shaped slip rate, which is constant along the fault. The rise time amounts to 0.48 seconds (Heaton 1990). The seismic moment M_0 was computed by a scaling relation (Hanks and Kanamori 1979; Kanamori 1977) assuming the local magnitude $M_L = 6.2$ to scale like the moment magnitude M_W (Ambraseys 2001) and amounts to $2.24 \cdot 10^{18}$ Nm. Further modeling parameters are compiled in Table 3.

Table 3. Earthquake modeling parameters

Strike (°)	Dip (°)	Rake (°)	Max. frequency (Hz)	Δt (s)	Δx (m)
0	90	0	1.5	0.015	200

Δt temporal discretization, Δx spatial discetization

4. Discussion

4.1 Peak Ground Velocities

Peak ground velocities have been determined from the synthetic time-histories simulated for 275000 sites at the surface. Areal plots of averaged horizontal peak ground velocity (PGV) are shown in Fig. 4. The average is computed as $\sqrt{(x^2 + y^2)}$ where x is the horizontal component perpendicular to the strike of the fault, and y the component parallel to the strike of the fault, respectively.

We will start the discussion with scenario 3, which shows the strongest ground motion and move then on to the other simulations. The largest overall peak ground velocity for scenario 3 (10.3 m/s) is observed within the basin close to the epicenter. Along the rupture, energy is radiated mainly within the basin causing the peak values within the basin to be more than three times as large as symmetrically outside of it. This is due to the fact that the displacement amplitude of a seismic point source scales with the inverse third power of the shear wave velocity. Thus, an event produces higher ground motion if the source is located in a low-velocity medium. For this simulation, the major part of the energy is trapped within the basin where the peak ground velocity takes an average value of 1.35 m/s. Compared to the overall mean PGV for this scenario which amounts to 0.39 m/s, this value is rather large. The fact that most of the energy is confined within the basin depicts its wave-guide behavior. However, the distribution of large PGV's for this scenario does not coincide with the areas of large intensities in Fig. 2.

A slightly smaller maximum PGV is found for scenario 1. It amounts to 8.1 m/s and is located 300 m to the northeast of the epicenter. The mean PGV's however, are considerably lower for this simulation compared to the former one. Within the basin the mean PGV amounts to only 0.28 m/s while the overall mean PGV has a value of 0.11 m/s. This feature can be explained by the fact that for scenario 1 most of the large PGV's are found in the vicinity of the epicenter. The mean PGV computed in a circle with radius 3 km around the epicenter amounts to 2.60 km/s compared to 2.53 km/s for scenario 3. Furthermore, due to rupture directivity, wave amplitudes north of the epicenter are larger than south of it for scenario 1. The largest increase is found just north of the northeastern and northwestern basin edge. We compare the values to PGV's at the sites symmetrically to those south of the epicenter and compute ratios of 1.3 and 1.6, respectively. However, the distribution of large PGV's is still remarkably different from the intensity pattern shown in Fig. 2.

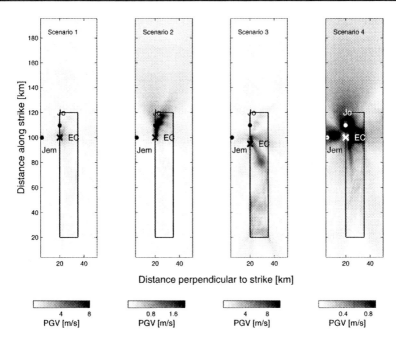

Fig. 4. From left to right: Averaged horizontal peak ground velocity (PGV) at the surface for scenarios 1-4. The basin is outlined by a black rectangle. The epicenter (*EC*) is displayed by a cross. The location of the cities of Jericho (*Jo*) and Jerusalem (*Jem*) are indicated by dots.

The smallest mean PGV's are determined for scenario 2. For the stations within the basin the value amounts to 0.15 m/s while the mean PGV for all sites takes a value of 0.09 m/s. The maximum PGV for this scenario is smaller than for the two scenarios mentioned above and amounts to 2.02 m/s. It is located 1 km north of the epicenter and 400 m east of the fault within the basin. For this simulation the effect of rupture directivity becomes most obvious: The largest PGV's are found north of the epicenter within the basin where the average amounts to 0.74. Additionally, waves reflected at the basin edges interfere with later phases, thus contributing to high values of PGV north of the epicenter. Just outside the basin, at its northeastern and northwestern corner, the PGV's amount to 0.17 and 0.39, corresponding to an increase of factors of 5.7 and 19.5 respectively, when compared to PGV's .At the same distance to the south of the epicenter.

For scenario 4, mean PGV's are only slightly larger than for scenario 2, both, within the basin (0.22 m/s) and when determined for the whole area (0.13 m/s). Here, the small values might be caused by the fact that the fault plane as a whole is located within the bedrock where the shear wave velocity is considerably higher than in the basin. Additionally, the distance from

the source to the free surface is larger in this simulation leading to decreased wave amplitudes due to geometrical spreading. Increased PGV's
are found in three elliptically shaped regions north, east, and west of the
source. Of those, the patch north of the epicenter contains the overall maximum PGV that amounts to 0.92 m/s and is found just above the northern
tip of the fault, 12 km north of the epicenter. The maximum PGV within
the patch that covers the basin takes a value of 0.86 m/s while the
maximum PGV in the patch west of the epicenter and thus outside the basin amounts to 0.55 km/s. North of the basin, high PGV's are found cen-
trally located north of the basin while at the southern end the seismic energy leaves the basin at its eastern and western corner forming a shadow
straight south of the basin. A diagonal pattern of high PGV's observed in
the center might be explained by reflections of waves within basin. The
overall pattern shows no similarities to the distribution of large intensities
in Fig. 2.

4.2 Spectral Accelerations

Spectral accelerations have been computed from the synthetic ground accelerations at every site for frequencies of 0.5 Hz, 1 Hz, and 1.5 Hz in order to get a better estimate of the frequency content of the seismic signal.

For all four scenarios, it was found that the highest overall values of
spectral accelerations were found at 1.5 Hz. Peak spectral accelerations are
listed in Table 4. The locations of the peak values coincide with those of
the maximum PGV's for all four scenarios thus being located in the proximity of the epicenter. These results are not in good agreement with the
observations made by Sieberg who reports on minor damage to the city of
Jericho itself.

Table 4. Peak spectral accelerations (cm/s^2)

	Scenario 1	Scenario 2	Scenario 3	Scenario 4
0.5 Hz	385	799	695	328
1 Hz	801	1585	1594	787
1.5 Hz	1126	2203	2594	1502

Areal plots of spectral accelerations are shown in Figs. 5, 6 and 7. We
will not discuss them in detail, but will concentrate our analysis on the area
north of the DSB where Sieberg observed considerable damage during the
earthquake of 1927.

Scenario 1: The pattern for the spectral acceleration at 0.5 Hz shows a local maximum 20 km north of the epicenter. It is located approximately 10 km south of the area of maximum intensity determined by Sieberg. Due to its small size, however, it cannot account well for the large macroseismic intensities. The spectral accelerations at 1 Hz and 1.5 Hz do not show increased values in this area. Slightly larger spectral accelerations are found in both cases approximately 20 km north of the epicenter. No significant increase in the spectral acceleration was found in the two areas approximately 100 km to the northeast and to the northwest of the epicenter where Sieberg found macroseismic intensities of IX. The computations of scenario 1 predict a minor maximum 5 km north of the center of the basin for spectral accelerations at 1 Hz. This maximum does not match that of the intensity map.

Scenario 2: The best qualitatively fit is observed for the computation of scenario 2. For all three frequencies considered two lobes of increased amplitudes develop, one being located to the northeast of the basin, the other one to its northwest. While for 0.5 Hz the lobe to the northwest of the basin is more pronounced, the behavior is reversed when spectral accelerations at frequencies of 1 Hz and 1.5 Hz are analyzed. Two local maxima in the spectral accelerations form approximately 65 km to the northeast and northwest of the basin for the spectral acceleration at 1.5 Hz with values of 187 cm/s^2 and 181 cm/s^2.

Scenario 3: The distribution of spectral accelerations for scenario 3 is dominated by large values within the basin, despite additional increased values north of the epicenter. For 1 Hz and 1.5 Hz the values increase to the northeast and the northwest of the basin. Local maxima outside the basin, however, cannot be found for this scenario.

Scenario 4: This scenario does not reproduce the two patches of increased amplitudes north of the basin for any of the frequencies considered. The values computed by scenario 4 predict large spectral accelerations 40 to 60 km north of the epicenter that differ from the distribution pattern of intensities of Sieberg's map.

Despite slight differences, the spectral accelerations predicted by scenario 2 show the best agreement with Sieberg's intensity map. Since we are well aware that different patterns may be observed for higher frequencies, modeling high frequency ground motion (e.g. frequencies up to 4 Hz) within the DSB area remains an important challenge for future research. Furthermore, local site effects may enhance ground motion amplitudes considerably. Therefore, site conditions and small-scale features should be included when modeling higher frequencies.

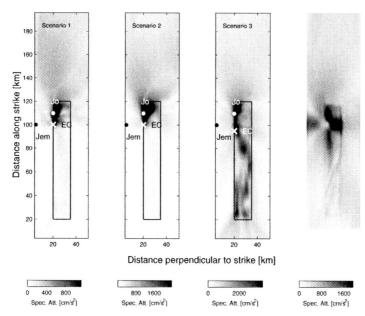

Fig. 5. From left to right: Spectral accelerations (0.5 Hz) at the surface for scenarios 1-4. The basin is outlined by a black rectangle. The epicenter (*EC*) is displayed by a cross. The location of the cities of Jericho (*Jo*) and Jerusalem (*Jem*) are indicated by dots.

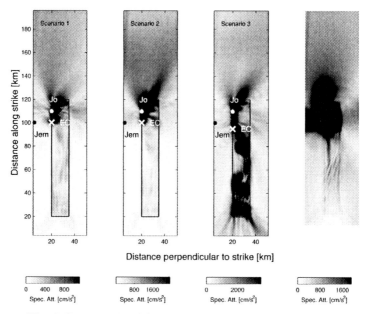

Fig. 6. Same as Fig. 5, but for spectral accelerations at 1 Hz.

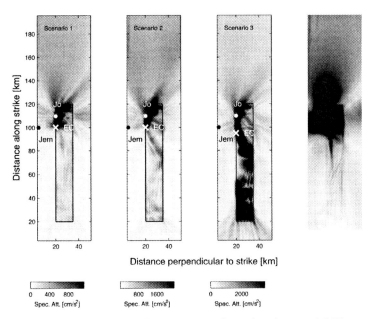

Fig. 7. Same as Fig. 5, but for spectral accelerations at 1.5 Hz.

5. Summary

The macroseismic intensity distribution for the 1927 Jericho earthquake is significantly asymmetric with large (> IX) intensities confined to three patches to the north of the epicenter while the damage in the epicentral region is rather limited (Sieberg 1932). We have tried to explain the distribution of intensities mapped qualitatively (Sieberg, 1932) by 3D FD modeling of wave propagation of four possible earthquake scenarios. The four scenarios differ by their source and rupture parameters such as the hypocenter and the fault plane location or the rupture velocity. Peak ground velocities and spectral accelerations have been computed from time-histories at the surface of our model and then compared to the intensity map. The best overall fit is obtained by scenario 2, a unilateral rupture on a vertical strike-slip fault. The rupture starts at the southern end of the fault and propagates with a velocity of 90% of the local shear wave speed, a velocity that has been purposely set to be larger than what is generally proposed in the literature (Geller 1976). This simulation can however give an overall better fit to the intensity pattern than scenario 1 where the rupture velocity was set to 70% of the local shear wave speed. Further, the maximum PGV computed for scenario 2 (2.02 m/s) seems more realistic than the maximum PGV's computed for scenarios 1 and 3 (8.1 m/s and 10.3 m/s, respectively). Due to the fact that for scenario 2 the rupture starts at the southern end of the fault plane, rupture directivity effects, giving rise to enhanced amplitudes north of the rupture, play a crucial role for this simulation. A comparison to scenario 3 shows that in the case of a bilateral rupture most of the energy is trapped within the basin making it more difficult to reproduce the intensity pattern. The predictions of scenario 2 were found to match Sieberg's intensity maps best, in particular north of the Dead Sea. In addition, its fit to the intensity map even increases as the frequency level increases. The computation of spectral accelerations at frequencies above 1.5 Hz could possibly reinforce that trend. It is also expected that the integration of local site conditions and mountain topography would significantly influence the predictions when modeling higher frequencies. This might play a crucial role for the area located 50 km to the northwest of the basin, just to the east of the Ephraim mountains, where Sieberg found the maximum intensities.

Acknowledgements

The authors wish to thank Kim B. Olsen for making available his FD-code. We are thankful to K. Fuchs and to H. Igel for their careful reviews of the manuscript, their helpful comments and suggestions. The study has been supported by the Collaborative Research Center 461 'Strong Earthquakes: A Challenge for Geosciences and Civil Engineering' of Karlsruhe University, funded by Deutsche Forschungsgemeinschaft (DFG).

References

Al-Zoubi A, Shulman H, Ben-Avraham Z (2002) Seismic reflection profiles across the southern Dead Sea basin. Tectonophysics, in press
Ambraseys NN (2001) Reassessment of earthquakes, 1900-1999, in the eastern Mediterranean and middle east. Geophys J Int 145:471-485
Ambraseys NN, Melville CP, Adams RD (1994) The seismicity of Egypt, Arabia, and the Dead Sea: a historical review. Cambridge University Press, Cambridge
Amiran DHK, Arieh E, Turcotte T (1994) Earthquakes in Israel and adjacent areas: macroseismic observations since 100 BCE. Isr Exploration J 44:260-305
Ben-Menahem A (1979) Earthquake catalogue for the middle east (92 BC – 1980 AD). Boll Geofis Teor Appl 21:245-310
Ben-Menahem A (1991) Four thousand years of seismicity along the Dead Sea Rift. J Geophys Res 96:20195-20216

Ben-Avraham, Z (1997) Geophysical framework of the Dead Sea: structure and tectonics. In: Niemi, TM, Ben-Avraham, Z, Gat, JR (eds) The Dead Sea: The Lake and its settings. Oxford University Press, New York, pp 22-35

Fukuyama E, Olsen KB (2002) A condition for super-shear rupture propagation in a heterogeneous stress field. Pure Appl Geophys (in press)

Garfunkel Z, Ben-Avraham Z (1996) The structure of the Dead Sea Basin. Tectonophysics 266:155-176

Geller RJ (1976) Scaling relations for earthquake source parameters and magnitudes. Bull Seis Soc Am 66:1501-1523

Ginzburg, A, Ben-Avraham Z (1997) A seismic refraction study of the north basin of the Dead Sea. Geophys Res Lett 24:2063-2066

Gottschämmer E, Olsen KB (2001) Accuracy of the explicit planar free-surface boundary condition implemented in a fourth-order staggered-grid velocity-stress finite-difference scheme. Bull Seis Soc Am 91:617-623

Gottschämmer E, Olsen KB, Wenzel F (2000) Finite-difference modeling of earthquakes in the Upper Rhinegraben, Germany. Eos Trans 81:F827

Gottschämmer E, Wenzel F, Wust-Bloch H, Ben-Avraham Z (2002) Earthquake modeling in the Dead Sea Basin. Geophys Res Lett, in press

Hanks TC, Kanamori H (1979) A moment magnitude scale. J Geophys Res 84:2348-2350

Heaton T (1990) Evidence for and implications of self-healing pulses of slip in earthquake rupture. Phys Earth Planet Inter 64:1-20

Kanamori H (1977) The energy release in great earthquakes. J Geophys Res 82:2981-2987

Nur A, Ron H (1996) And the walls came tumbling down: earthquake history in the holyland. In: Stirros S, Jones RE (eds) Archeoseismology. Institute of Geology and Mineral Exploration, Athens, pp 75-85

Olsen KB (1994) Simulation of three-dimensional wave propagation in the Salt Lake Basin. Ph.D. thesis, University of Utah, Salt Lake City

Olsen KB (2000) Site amplification in the Los Angeles Basin from 3D modeling of ground motion. Bull Seis Soc Am 90:S77-S94

Sato T, Graves RW, Somerville RG (1999) Three-dimensional finite-difference simulations of long-period strong motions in the Tokyo metropolitan area during the 1990 Odawara Earthquake (M_J 5.1) and the Great 1923 Kanto Earthquake (M_S 8.2) in Japan. Bull Seis Soc Am 89:579-607

Shapira A, Ron H, Nur A (1993) A new estimate for the epicenter of the Jericho earthquake of 11 July 1927. Israel J Earth Sci 42:93-96

Sieberg A (1932) Untersuchungen über Erdbeben und Bruchschollenbau im östlichen Mittelmeer. Fischer, Jena

Wells DL, Coppersmith KJ (1994) New empirical relations among magnitude, rupture length, rupture width, rupture area, and surface displacement. Bull Seism Soc Am 84:974-1002

Wust-Bloch GH (2002) The active Dead Sea Rift fault zone: a seismic waveguide. Stephan Mueller Special Publication Series (in press)

Wust-Bloch, GH, Wachs D (2000) Seismic triggering of unstable slopes in northern Israel. Israel J Earth Sci 49:103-109

Nonlinear Seismology – The Seismology of the XXI Century

Gh. Marmureanu, M. Misicu, C.O. Cioflan, F.St. Balan, and B.F. Apostol

National Institute for Earth Physics, Bucharest, Romania

Abstract

Nonlinear effects in ground motion during large earthquakes have been a long controversial issue between seismologists and geotechnical engineers. Laboratory tests made by using resonant columns consistently show the reduction in shear modulus (G) and increase in damping ratio (D) with increasing shear strain (γ), i.e., $G=G(\gamma)$, respectively, $D=D(\gamma)$. Therefore nonlinear viscoelastic constitutive laws are required. The seismological detection of the nonlinear site effects requires a simultaneous understanding of the effects of earthquake source, propagation path and local geological site conditions. The difficulty for seismologists in assessing the nonlinear site effects is due to the fact that these are overshadowed by the overall patterns of shock generation and propagation. In order to bring evidence for nonlinear effects we have introduced the spectral amplification factor (SAF) as the ratio between the maximum values of the spectral absolute acceleration (Sa), relative velocity (Sv), and relative displacement (Sd) to the corresponding peak values obtained from the processed strong motion records. The main evidence of nonlinearity observed for the thick Romanian Plain Quaternary sediments is a systematic decrease in the variability of peak ground acceleration for increasing earthquake magnitudes.

1. Introduction

In 1983 a conference of seismologists and geotechnical engineers was convened by two federal agencies - the United States Nuclear Regulatory Commission and the United States Geological Survey - to examine the state of the art in evaluating the effects of local site conditions on strong ground motion (Hays, 1983). The topic that created the most spirited discussion centered on the issue of the nonlinearity of the site response. The conference has not reached a consensus on the nonlinear effects in soil response (Reiter, 1990). Simply put, nonlinearity is that phenomenon that allows for changes in soil properties and therefore changes in soil response as the level of the ground motion increases.

Nonlinearity in soil response is also called strain dependence, because the strain the soil goes through during an earthquake increases with the level of stress or ground motion. The nonlinear effects in seismic motion are discussed extensively in recent papers (e.g., Yokohama Symposium, 1998). Direct observation of such nonlinear effects is claimed to have been read on accelerograms ´(Archuleta, 1998), and nonlinear characteristics of soil response are described in some case studies (Ni et al, 1997). It is also worth noting that numerical simulations of nonlinear components of seismic motion were performed (Pavlenko, 2001), and the state-of-the-art of the issues that are presumably related to, or caused by nonlinear behavior was presented recently (Field, 1998). In a recent paper, entitled 'Nonlinear Soil Response – A reality?' (Beresnev and Wen, 1996), the decrease of the local site amplification is pointed out as one of the nonlinear effects. It is this particular point on which we focus our attention in the present paper, by discussing nonlinear effects in spectral amplification factors (SAFs) with application to the Romanian Plain Quaternary Sediments.

2. Geotechnical Evidence for Nonlinear Soil Behaviour

It is well established in geotechnical engineering that soil response is nonlinear beyond a certain level of deformation. Laboratory tests made on Hardin and Drnevich resonant columns, such as those shown in Figures 1, 2a and 2b, consistently show the reduction in shear modulus (G) and increase in damping ratio (D) with increasing shear strain (γ) above 10^{-4} %, i.e. these parameters are functions of shear strain, $G=G(\gamma)$ and $D=D(\gamma)$. As we can see in Figures 2a and 2b, the reduction in shear modulus for limestone and gritstone is only 0.80, and respectively 0.72 for a high shear strain level $\gamma= 0.01$. However, for sand with gravel these parameters may acquire higher values, as one can see in Fig 1.

In analyzing the seismic response of ground or soil structures by employing the theory of wave propagation or the finite element principle, it is of uttermost importance to represent the cyclic behavior of soils by a material model which correlates the shear stress and shear strain. Modeling of soil behavior under cyclic or random loading conditions must be made so that the model can reproduce the deformation characteristics in the range of strains under consideration.

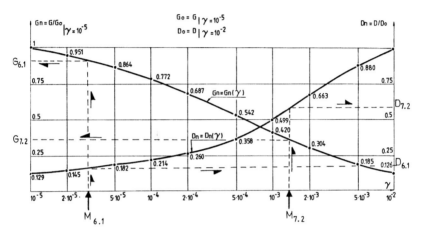

Fig. 1. Shear modulus G and damping ratio D vs strain γ for sand with gravel from Drnevich resonant column

When soil or rock behavior is expected to remain in the range of the small strain, the use of an elastic model is appropriated, and the shear modulus is a key parameter to describe properly the soil or rock behavior. This may be adequate for the propagation path of waves from source to the fundament of the site, or for the local special geological site conditions (limestones, gritstones etc.). When a given problem is associated with the medium range of strain, approximately below the level of 10^{-3} the soil behavior becomes elasto-plastic and the shear modulus tends to decrease on increasing the shear strain (Figures 1, 2a and 2b). The damping ratio can be used in this case to describe the energy absorbing capacity of soils.

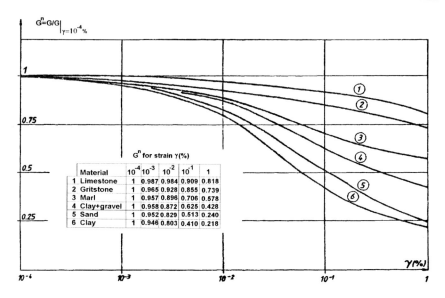

Fig. 2a. Nonlinear relationship between shear modulus G and shear strain γ for various soils, from Hardin-Drnevich resonant column measurements

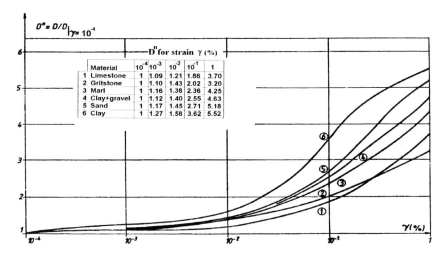

Fig. 2b. Nonlinear relationship between damping ratio D and shear strain γ for various soils, from Hardin-Drnevich resonant column measurements

If the strain level concerned is still small enough not to cause any progressive change in soil properties, the shear modulus and damping ratio do not change with the progression of cycles in load application. This kind of behavior is called nondegraded hysteresis type and such steady-stage soil characteristics can be represented to a reasonable degree of accuracy by use of the linear viscoelastic theory. The shear modulus (G) and damping ratio (D) determined as functions of shear strain (γ) are the essential parameters to represent soil properties in this medium strain range. The apparently pervasive nonlinearity in the amplification factor and strong nonlinear dependence of the shear modulus and damping ratio with shear strains induced by strong earthquakes at sediment sites is a serious challenge to seismologists, among whom the principle of linear elastic theory is widely accepted for strong motion prediction, as involved, for instance, by the so-called empirical Green's function method.

The main question is how the nonlinear deformation process manifests itself in seismological observation and what particular seismic effects can be associated, or traced back, to the nonlinear soil properties. A convincing direct evidence for nonlinear effects in strong seismic motion as gathered from seismological observations is still wanted (Aki et al, 1980).

3. Constitutive and Dynamic Equations.
Approximate Solutions of Wave Equations

The nonlinear behavior of soil layers at earth surface may considerably alter the seismic maps that represent the isoseistic earthquake lines, as well as the linear theory results, at least on a local scale. According to the previously developed approximation scheme (Doyle and Ericson, 1959; Signorini, 1936), the nonlinear effects may be determined successively up to higher order. Herein we do not consider very high nonlinearities, so that we restrict ourselves to the second-order approximation level (Misicu, 2002a, Misicu et al, 2002b). A more refined analysis concerning nonlinearities of constitutive laws has been performd recently, and its results on fractal, plastic and damaged media will be taken into account in a forthcoming publication.

The constitutive and dynamic equations read:

$$\sigma_{ij} = \delta_{ij}\lambda(\theta)\theta + 2\mu(\gamma)\varepsilon_{ij}, \sigma_{ij,j} = \rho\ddot{u}_i \qquad (3.1)$$

where $\theta = \varepsilon_{ii}$ is the dilatation, $\varepsilon_{ij} = \dfrac{1}{2}(u_{i,j} + u_{j,i} + u_{k,i}u_{k,j})$ is the strain tensor, $i,j=1,2,3$), $\gamma = \sqrt{\varepsilon^D_{ij}\varepsilon^D_{ij}}$ is the second-order invariant of the strain deviator, $\varepsilon^D_{ij} = \varepsilon_{ij} - \dfrac{1}{3}\delta_{ij}\theta$, ρ is the density and σ_{ij} is the stress tensor.

The coefficients λ and μ are assumed to be polynomials of dilatation and second order strain invariant,

$$\lambda(\theta) = \sum_{n=0}^{2} \lambda_n \theta^n , \mu(\gamma) = \sum_{n=0}^{2} \mu_n \gamma^n \qquad (3.3)$$

A more general Ansatz would consist in the assumption that these moduli are function of both invariants θ, γ. As it is well-known the distortion function and the rotation are given by:

$$u_{i,j} = \varepsilon_{ij} - \omega_{ij} - \frac{1}{2} u_{k,i} u_{k,j} \quad \text{and} \quad \omega_{ij} = \frac{1}{2}(u_{j,i} - u_{i,j}) \qquad (3.4)$$

Introducing the stress tensor into the equation of motion given by (3.1) one obtains the differential equation of motion:

$$(\lambda + \mu)\theta_{,i} + \mu(\Delta u_i + u_{k,i}\Delta u_k + u_{k,ij}u_{k,j}) + \lambda_{,i}\theta + \qquad (3.5)$$

$$\mu_{,j}(u_{i,j} + u_{j,i} + u_{k,i}u_{k,j}) = \rho\ddot{u}_i$$

Which is a wave equation with nonlinear terms.

3.1 Approximate Solutions of Wave Equations

Following the approximation schemes introduced by Doyle et al 1956) we consider the following expansions (Misicu et al, 2002/2):

$$u_i = \sum_{n=1,2,...} e^n u_{ni} \qquad (3.6)$$

e being a small parameter (<1). Consequently, the dilatation and rotations may be expanded as follows:

$$\theta = \sum_{n=1,2..} e^n \theta_n , \theta_n = u_{ni,i} , \omega_{nij} = (u_{nj,i} - u_{ni,j})/2 , \gamma = \sum_{n=1,2...} e^n \gamma_n \qquad (3.7)$$

Making use of these expansions equation (3.5) leads to dilational and rotational wave equations

$$(\lambda_0 + 2\mu_0)\Delta\theta_1 = \rho\ddot{\theta}_1 \qquad (3.8)$$

$$(\lambda_0 + 2\mu_0)\Delta\theta_2 + \Theta_2 = \rho\ddot{u}_{2i} , (\lambda_0 + 2\mu_0)\Delta\theta_3 + \Theta_3 = \rho\ddot{\theta}_3 \qquad \text{and}$$

$$\mu_0\Delta\omega_{1ij} = \rho\ddot{\omega}_{1ij} , \mu_0\Delta\omega_{2ij} + \Omega_{2ij} = \rho\ddot{\omega}_{2ij} , \mu_0\Delta\omega_{3ij} + \Omega_{3ij} = \rho\ddot{\omega}_{3ij} ,... \qquad (3.9)$$

where $\Theta_2 = E_{2i,i} , \Theta_3 = E_{3i,i}$

$$\Omega_{2ij} = \frac{1}{2}(E_{2j,i} - E_{2i,j}), \Omega_{3ij} = \frac{1}{2}(E_{3j,i} - E_{3i,j}) \tag{3.10}$$

$$E_{2i} = (\lambda_1\theta_1 + \mu_1\gamma_1)\theta_{1,i} + \mu_0(u_{1k,i}\Delta u_{1k} + u_{1k,ij}u_{1k,j}) + \mu_1\gamma_1\Delta u_{1i}$$
$$+ \lambda_1\theta_{1,i}\theta_1 + \mu_1\gamma_{1,j}(u_{1i,j} + u_{1j,i})$$

$$E_{3i} = (\lambda_1\theta_1 + \mu_1\gamma_1)\theta_{2,i} + (\lambda_1\theta_2 + \lambda_2\theta_1^2 + \mu_1\gamma_2 + \mu_2\gamma_1^2)\theta_{1,i} + \tag{3.11}$$
$$\mu_0(u_{1k,i}\Delta u_{2k} + u_{2k,i}\Delta u_1 + u_{1k,ij}u_{2k,j} + u_{2k,ij}u_{1k,j}) +$$
$$\mu_1\gamma_1(\Delta u_{2i} + u_{1k,i}\Delta u_{1k} + u_{1k,ij}\Delta u_{1k,j}) + (\mu_1\gamma_2 + \mu_2\gamma_1^2)\Delta u_{1i} +$$
$$\lambda_1\theta_{1,i}\theta_2 + \lambda\theta_{2,i}\theta_1 + \mu_1\gamma_{1,j}(u_{2i,j} + u_{2j,i} + u_{1k,i}u_{1k,j}) +$$
$$\mu_2\gamma_{2,j}(u_{1i,j} + u_{1j,i})$$

And where $\gamma_1 = \gamma_{11}, \gamma_2 = \gamma_{11}(\gamma_{22}/\gamma_{11})^2, ...,$

3.2 Nonlinear SH Plane Waves Incident to a Solid-Solid Interface

Making use of the linear solutions of the first equation given by Eq. (3.9), which correspond to *SH* waves satisfying the conditions $u_{11} = u_{13} = 0, u_{12} \neq 0$, we get $\omega_{123} = -u_{12,3}/2, \omega_{112} = u_{12,1}/2$, $\omega_{131} = 0$ so that the equation reduces to: $\Delta u_{12} - \beta^{-2}\ddot{u}_{12} = 0$ where $\beta^2 = \mu_0/\rho$. This equation admits the solution:

$$u_{12} = e^{i\xi}[u''e^{i\varsigma} + u'e^{-i\varsigma}], (\xi = \omega t - kx\sin f, \varsigma = kz\cos f) \tag{3.12}$$

Here $k = \omega/\beta$, f is the incidence angle between the normal to the waves front and the vertical axis and u' and u'' are the coefficients of the linear solutions. We denote by $c = \beta/\sin f$ the horizontal phase velocity. Straightforward calculations lead to the second-order approximation (Misicu 2002/1, Misicu et al 2002/2)

$$\gamma = \frac{ik}{\sqrt{2}}e^{i\xi}\left(au''e^{i\varsigma} + bu'e^{-i\varsigma}\right) \tag{3.13}$$

Where the coefficients a and b account for the character of the incident wave, namely $a=1$, $b=0$ for the ascendent waves and $a=0$, $b=1$ for the descendent waves.

Similarly, the rotational terms in Eq. (3.10) become:

$$\Omega_{212} = -\mu_1 k^4 e^{2ik\xi} \sin f[au''^2 e^{2i\varsigma} + bu'^2 e^{-2i\varsigma} + (a+b)u''u'\sin^2 f]$$

$$\Omega_{223} = -\mu_1 k^4 e^{2i\varsigma} \cos f[au''^2 e^{2i\varsigma} - bu'^2 e^{-2i\varsigma}], \quad \Omega_{231} = 0 \qquad (3.14)$$

and the rotation wave equation reads:

$$\mu_0 \Delta\omega_{212} + \Omega_{212} = \rho\ddot{\omega}_{212}, \mu_0\Delta\omega_{223} + \Omega_{223} = \rho\ddot{\omega}_{213}, \qquad (3.15)$$
$$\mu_0\Delta\omega_{231} = \rho\ddot{\omega}_{231}$$

The last equation is not relevant, because it is reducible to a similar one obtained from Eq. (3.9). We consider plane waves solutions:

$$u_{22} = e^{2i\varsigma}[u_2 + e^{2i\varsigma}(u''_2 + w''_2 \xi + \alpha''_2 \zeta) + e^{-2i\varsigma}(u'_2 + w'_2 \xi + \alpha'_2 \zeta)] \qquad (3.16)$$

where the coefficients α_2'', α_2', u_2'', u_2' have to be determined from the boundary conditions. By introducing (3.16) into Eq. (3.11) we get the conditions:

$$u_2 = \frac{\mu_1}{\mu_0}\cdot\frac{k(a+b)u''u'tg^2 f}{4}, \quad w''_2 = -\frac{\mu_1}{\mu_0}\cdot\frac{kau''^2}{4} - \alpha_2''ctg^2 f, \qquad (3.17)$$

$$w'_2 = -\frac{\mu_1}{\mu_0}\cdot\frac{kbu'^2}{4} + \alpha_2'ctg^2 f$$

We note that the wave amplitudes depend upon the ratio of the first-nonlinear contribution of the transversal modulus to its linear part, as well as on the nonlinear contribution to the displacements. It is obvious that Eq. (3.16) provides us with amplification quotients of the amplitudes. More specifically these are given by:

$$A_2'' = 1 \pm \left|\frac{u_{22}}{u''}\right|; \quad A_2' = 1 \pm \left|\frac{u_{22}}{u'}\right| \quad \text{,etc.} \qquad (3.18)$$

where $\pm$ signs correspond to ascendent or descendent waves, respectively.

Eq. (3.18) represents the amplification of the wave front in the second-order approximation with respect to the first order. If the waves occur in a superficial layer, then, by taking into account the reflections, we have to double the amplifications.

These results may be applied to particular cases. For a simple layer with free surface, thickness d and receiving for the underground only front waves with period ω, we have the boundary conditions:

$$\sigma_{123} + \sigma_{223} = 0, (z = 0); u_{22} = 0, (z = d) \text{ for the following} \qquad (3.19)$$

constitutive equations $\sigma_{123} = \mu_0 u_{12,3}, \quad \sigma_{223} = \mu_0 u_{22,3} + \mu_1 u_{12,3}$

Finally we obtain the particular expresions of solutions and coefficients are given by:

$$u'' = u', \qquad (3.20)$$

$$u'_2 = M[A(1-2i\theta e^{4i\theta}) - B(1+2i\theta)/ctg^2 f - 2iC e^{2i\theta}]/2i[1+(1+2\theta)e^{4i\theta}]$$

$$u'_2 = -MA[1 + 2\theta(1+i)e^{4i\theta}]/2i[1 + 2(1+i)\theta)e^{4i\theta}]ctg^2 f$$

$$\alpha''_2 = -MA/ctg^2 f, \alpha'_2 = MB/ctg^2 f, \ w''_2 = w'_2 = 0$$

$$A = au''^2, B = bu'^2, C = (a+b)u''u'tg^2 f/i$$

Calculation can be simplified by assuming ascendent waves for $B=C=0$; we get

$$a_{22} = MAe^{2i\xi}(2i\theta - 1)(e^{2i(2\theta - \zeta)} - e^{2i\zeta})/2i\cos^2 f(1 + 4i\theta) \qquad (3.21)$$

$$u_{12} = 2u''\cos\theta, \qquad \theta = kd\cos f$$

Further on, we introduce the notation $l'' = au'', L = \dfrac{\beta T ctgf}{\sqrt{2\pi}} \cdot \sqrt{\dfrac{l''}{d}}, T = 2\pi/\omega, m = \mu_1/\mu_0, \Lambda'' = l''/L$, which allows Eq. (3.18) to be rewritten as: $A''_2 = 1 + m\Lambda''$. In Fig.3 the amplitude coefficient A''_2 is plotted vs the characteristic length L for given values of $n = au''$. Similarly, A''_2 is plotted in Fig. 4 vs m for given values of the ratio Λ''. One can see in Fig.3 that amplification depends on the characteristic length L. Herein, the characteristic length L acquires rather small values, corresponding to small periods and propagation velocities (Misicu et al, 2002b). According to our solutions given by (3.16) one can say that nonlinearities imply indeed a frequency-dependent amplification for a layered crust, including the shift of the fundamental resonance towards a lower frequency and the appearance of higher-frequency resonances (doubling the frequency in our approximation). Such effects are usually pointed out in standard qualitative analysis of anharmonic corrections to wave equation. For a shear modulus of 100 daN/cm^2, there is an increasing in amplitude of about 30%. This corresponds to a maximum displacement of 30 cm and a period of oscilation of 1s for soils with density of 1800-2000 kg/m^3. Hence, the seismic effects may considerably increase (with a percentage of 30%) if the wave front is characterized by higher frequencies and longer travel times, as due to altered structure (large deformations) of earth, or to the superpositions of radiated waves from multiple sources. For March 4, 1977 Vrancea event the sources were located at relatively large distances, stretching over 60 km, and the reported shock sequences have induced wave fronts with different orientations and amplitudes during short time intervals. Usually, the standard design procedures disregard such instances of strong effects, or special peak effects.

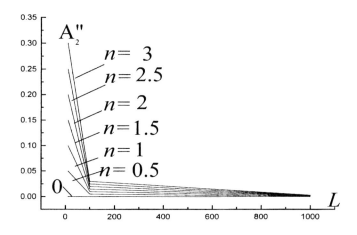

Fig. 3. Amplitude coefficient A''_2 vs the characteristic length L for various values of the parameter n

It is also worth noting that our amplification factors as given by (3.18) include directivity effects, through the dependence on the incident angle. Such effects have been pointed out previously in the March 4, 1977 Vrancea earthquake. The directivity curve indicates the superposition of principal wave fronts radiated from major seismic sources (Misicu et al, 1982, Cornea et al, 1980, 1981).

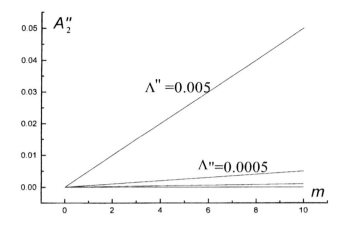

Fig. 4. Amplitude coefficient A''_2 vs ratio m for values of the parameter Λ''

4. Quantitative Evaluation of Nonlinearity

The model of linear elastic response of the earth has been almost universally used by seismologists to model teleseismic, weak and also strong earthquakes. For teleseismic and weak ground motions, there is no reason to doubt that this model is acceptable, but for strong ground motions, particularly when recorded on soils, the consequences of a nonlinear soil behavior have to be seriously considered. Ground motion characteristics, particularly the duration of strong shaking, can be affected by seismic energy being trapped in large sedimentary basins. For smaller earthquakes, the strains are smaller and we are on the left-hand side of Fig. 1; for strong earthquakes, the strains are larger and we are in the right-hand side of Fig. 1. Consequently the response of a system of nonlinear viscoelastic materials (clays, marls, sands etc.) subjected, for example to vertically travelling shear waves, is far away from being linear, thus generating large discrepancies with respect to a linear model. In this case, in the wave equation:

$$G\frac{\partial^2 u_2(x_1,t)}{\partial x_1^2} + D\frac{\partial^3 u_2(x_1,t)}{\partial t \partial x_1^2} = \rho\frac{\partial^2 u_2(x_1,t)}{\partial t^2} \qquad (4.1)$$

appears the D–term, where G is the dynamic torsional modulus function and D is the torsional damping function; these parameters are all functions of shear strains (γ), frequency (ω), confining pressure (σ), depth (h), temperature (t), void ratio (v_r) etc., that is: G = G($\gamma,\omega,\sigma,$h,t,v_r,…), D = D($\gamma,\omega,\sigma,$h,t,v_r,…).

In order to find the quantitative characteristics of the nonlinear soil behavior and nonliner site response, we introducved previously 'the spectral (seismic) amplification factor' (Marmureanu et al, 1995, 1996).

4.1 The Definition of the Spectral Amplification Factor (SAF)

If we have a one-degree-of-freedom linear oscillator of mass m, stifness k, damping ζ, subjected to base acceleration ä(t), or absolute displacement a(t) of the ground, the relative displacement x(t) of the mass *m* can be computed from Duhamel integral. For vanishing initial conditions, the expression for x(t) is given by:

$$x(t) = \frac{-1}{\omega\sqrt{1-\beta^2}}\int_0^t \ddot{a}(\tau)e^{-\beta\omega(t-\tau)}\sin\omega\sqrt{1-\beta^2}(t-\tau)d\tau \qquad (4.2)$$

where ω is the natural frequency and β represents the fraction of critical damping.

Similarly, the absolute acceleration ÿ(t)= $\ddot{x}(t)+\ddot{a}(t)$ is given by the expression:

$$\ddot{y}(t) = \omega\frac{1-2\beta^2}{\sqrt{1-\beta^2}}\int_0^t \ddot{a}(\tau)e^{-\beta\omega(t-\tau)}\sin\omega\sqrt{1-\beta^2}(t-\tau)d\tau + \qquad (4.3)$$

$$+2\omega\beta\int_0^t \ddot{a}(\tau)e^{-\beta\omega(t-\tau)}\cos\omega\sqrt{1-\beta^2}(t-\tau)d\tau$$

And the maximum absolute values of the quantities x(t), $\dot{x}(t)$ and ÿ(t) experienced during the earthquake response are commonly defined as follows:

$$S_d = |x(t)|_{max} \; ; \quad S_v = |\dot{x}(t)|_{max} \; ; \quad S_a = |\ddot{y}(t)|_{max} \qquad (4.4)$$

The quantities S_a, S_v and S_d as functions of the undamped natural period of vibration T=$2\pi/\omega$ for various fractions of critical damping (β=0%,2%,5%.10%,20% etc.) are called earthquake response spectra.The ratio of the maximum spectral values of S_a , S_v, S_d from response spectra for a fraction of critical damping (β) to peak values of $\ddot{y}(t)$, $\dot{x}(t)$, and, respectively, x(t), as obtained from the processed acceleration records, is called the spectral amplification factor for absolute acceleration(SAF_a), relative velocity(SAF_v), and, respectively, relative displacement (SAF_d):

$$SAF_a = \frac{S_a^{max}}{a_{max}}; \quad SAF_v = \frac{S_v^{max}}{v_{max}}; \quad SAF_d = \frac{S_d^{max}}{d_{max}} \qquad (4.5)$$

Where

$$a_{max} = \ddot{y}(t)_{max}; \quad v_{max} = \dot{x}(t)_{max} \quad \text{and} \quad d_{max} = x(t) \qquad (4.6)$$

In Figure 5 we can see the spectral amplification factors and the efect of the nonlinearity (grey area) at Cernavoda site for the strong Vrancea earthquake of August 30, 1986 ($M_S = 7.0$). One can see that the SAF coefficient is usefull to describe the nonlinearity fenomena at a given site, being indirectly dependent on the frequency content of the earthquake.

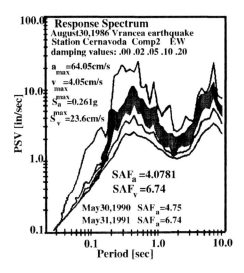

Fig. 5. SAF and the effect of nonlinearity (grey area) at Cernavoda site for the August 30, 1986 Vrancea earthquake ($M_S = 7.0$)

4.2 Spectral Amplification Factors for Recent Strong Vrancea Earthquakes

In Table 1 spectral amplification factors are given for absolute accelerations at 5% fraction of critical damping ($\beta = 5\%$) at five seismic stations: Bucharest-INCERC (soft soils, quaternary layers with a total thickness of 700 m); Bucharest-Magurele (sand, loess-350m); Cernavoda-NPP (marl, loess, limestone-270 m), Bacau (gravel, loess-20 m) and Iasi (loess, sand, clay, gravel 20÷60m) for the four recent Vrancea strong eartquakes: March, 4, 1977 (M_S=7.2); August, 30, 1986 (M_S =7.0); May, 30, 1990 (M_S =6.7) and May, 31, 1990 (M_S=6.1)

Table 1. Spectral amplification factors for four recent strong Vrancea earthquakes

Bucharest-Magurele Seismic Station

Earthquake	$a_{max}(g)$ recorded	$S^a_{max}(g)$ (β=5%)	SAF_a	c	$S_a^*(g)$ (β=5%)	$a^*(g)$	%
March 4,1977	-	-	-	-	-	-	-
August 30,1986	0.116	0.313	2.6982	1.3294	0.4160	0.1542	32.94
May 30,1990	0.092	0.330	3.5869	1.0000	3.5869	0.0920	-
May 31,1990	-	-	-	-	-	-	-

Bucharest-INCERC Seismic Station

Earthquake	$a_{max}(g)$ recorded	$S^a_{max}(g)$ (β=5%)	SAF_a	c	$S_a^*(g)$ (β=5%)	$a^*(g)$	%
March 4,1977	0.208	0.620	2.9807	1.1386	0.7059	0.2368	13,46
August 30,1986	0.096	0.254	2.6458	1.2827	0.3258	0.1231	28.27
May 30,1990	0.066	0.224	3.3939	1.0000	0.224	0.066	-
May 31,1990	-	-	-	-	-	-	-

Cernavoda NPPlant Seismic Station

Earthquake	$a_{max}(g)$ recorded	$S^a_{max}(g)$ (β=5%)	SAF_a	c	$S_a^*(g)$ (β=5%)	$a^*(g)$	%
March 4,1977	-	-	-	-	-	-	-
August 30,1986	0.064	0.261	4.0781	1.4185	0.3702	0.0907	41.87
May 30,1990	0.102	0.485	4.7549	1.2166	0.5900	0.1241	21.66
May 31,1990	0.0507	0.2933	5.7851	1.000	0.2933	0.0507	-

Bacau Seismic Station

Earthquake	$a_{max}(g)$ recorded	$S^a_{max}(g)$ ($\beta=5\%$)	SAF_a	c	$S^*_a(g)$ ($\beta=5\%$)	$a^*(g)$	%
March 4,1977	-	-	-	-	-	-	-
August 30,1986	0.0736	0.298	4.0489	1.4557	0.4338	0.107	45.57
May 30,1990	0.135	0.697	5.1629	1.1416	0.7957	0.1540	14.16
May 31,1990	0.0643	0.379	5.8942	1.000	0.3790	0.0643	-

Iasi Seismic Station

Earthquake	$a_{max}(g)$ recorded	$S^a_{max}(g)$ ($\beta=5\%$)	SAF_a	c	$S^*_a(g)$ ($\beta=5\%$)	$a^*(g)$	%
March 4,1977	-	-	-	-	-	-	-
August 30,1986	0.0695	0.230	3.3079	1.2895	0.2966	0.0896	28.95
May 30,1990	0.0991	0.403	4.0665	1.049	0.4230	0.1039	4.95
May 31,1990	0.0504	0.215	4.2658	1.000	0.2150	0.0504	-

The effects of the nonlinearity are clearly observed in the grey areas of the Figures 5 and 6. The coefficient c is the ration of the SAF for May 31, 1990 Vrancea earthquakes to the SAF corresponding to each previous earthquake; $S^*_a(g)$ and $a^*(g)$ are the maximum spectral acceleration, and, respectively, the maximum acceleration if the system would have a linear response to the fundamental frequency at the considered site.

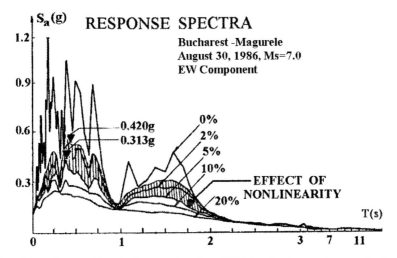

Fig. 6. Nonlinear effect in the acceleration SAF for Bucharest-Magurele site

Table 2. Median values of the spectral amplification factors for the three recent strong Vrancea earthquakes

Damping	August 30,1986 $M_s=7.0$		May 30,1990 $M_s=6.7$		May 31,1990 $M_s=6.1$	
$\beta\%$	SAF_a	SAF_v	SAF_a	SAF_v	SAF_a	SAF_v
2%	4.74	3.61	5.58	3.72	6.22	4.84
5%	3.26	2.69	3.63	2.95	4.16	3.48
10%	2.43	2.99	2.56	2.14	2.92	2.69
20%	1.78	1.50	1.82	1.58	2.13	1.86

For the Vrancea earthquake of May 31, 1990 ($M_s=6.1$) we may assume that the response is in the elastic range, so that we have the possibility to compare its effects to a linear model results. In addition, these spectral amplification factors are function of the earthquake magnitude. In Table 2 and Figure 7 we can see that there is a strong nonlinear dependence of the spectral amplification factors on earthquake magnitude (Marmureanu et al, 1996).

At the same seismic station, for instance Bacau, for horizontal components and 5% damping, the median values of the SAF for accelerations are: 3.94 for August 30,1986 Vrancea earthquake ($M_s=7.0$); 4.32 for May 30, 1990 ($M_s=6.7$) and 5.22 for May 31, 1990 ($M_s =6.1$). A characteristic of the nonlinearity is a systematic decrease in the variability of the peak ground acceleration with increasing earthquake magnitude. Spectral amplification factor decrease from 4.16 for Vrancea earthquake with magnitude Ms=6.1 to 3.26 for Vrancea strong earthquake with magitude Ms=7.00 (Table 2, damping 5%). The amplification factors decrease as the magnitude increase. This is consistent with data shown in Figure 7 and data from Tables 1 and 2, which confirm that the ground accelerations tend to decrease as earthquake magnitude increases. For example, if we mantain the same amplification factor (SAF=5.8942) as for relatively strong earthquake of May 31, 1990 with magnitude $M_s =6,1$ then at Bacau Seismic Sation for earthquake magnitude earthquake of May 30, 1990 ($M_s =6.7$) the peak acceleration has to be $a*_{max}=0.154g$ (+14.16%), while the recorded value was only, $a_{max}=0.135g$. Similarly, for Vrancea earthquake of August 30, 1986, the peak acceleration has to be $a*_{max}=0.107g$ (+45.57%), instead of the actual value of 0.0736 g recorded at Bacau Seismic Station (Table 1).

We must stress upon the fact that the SAF combines both resonant and nonlinear effects, so that it is rather difficult to distinguish between these effects from an overall analysis of these factors. For instance, a narrow-band response of a bedrock may exibit a higher SAF due to resonant effects in comparison with a broader-band response of a shallow soft soil, where the SAF is most likely due to nonlinear effects. Another nonlinear effect is also included in the saturation of the peak amplitudes on increasing earthquake magnitudes, a circumstance which enhances the difficulty of ascertaining the cause of the observed variation in the SAFs.

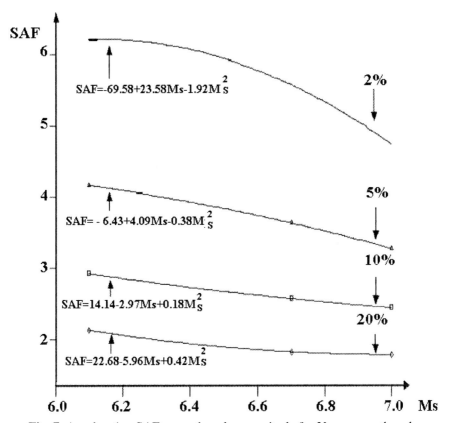

Fig. 7. Acceleration SAF vs earthquake magnitude for Vrancea earthquakes

5. Discussion

The significance of nonlinear-elastic soil response to strong earthquakes has been for long a contentious matter. On the one hand, soil samples behave nonlinearly in laboratory tests made on Hardin and Drnevich resonant columns at strains larger than 10^{-5} or 10^{-4}, a result that is standard in geo-technical research field. On the other hand, it is also routine in seismology to assume that soil amplification factors measured from weak motions apply to strong motions, i.e., effects of nonlinearity are complete neglected.

The model of linear elastic response of the earth has been almost universally used by seismologists to model teleseismic, weak and strong ground motions. Our opinion is that for teleseismic and weak ground motions, there is no reason to doubt that this model is acceptable, but for strong ground motions, particularly when recorded on thick Quaternary sediments, the consequences of nonlinear soil behavior have to be seriously considered.

In an attempt to understand the characteristics of the nonlinear soil behavior and the nonlinearity in the seismology, we have examined in the present paper the ways that nonlinearity would be expected to appear on strong motion records made on Romania teritory during the recent Vrancea earthquakes. In order to find the quantitative characteristics of nonlinear soil behavior and nonliner site response, we introduced the spectral (seismic) amplification factors (SAF) which are defined as the ratio between maximum spectral values of absolute acceleration (S_a), relative velocity (S_v) and displacement (S_d) from response spectra for a given fraction of critical damping (β) to the corresponding peak values of $\ddot{y}(t)$, $\dot{x}(t)$, and, respectively, x(t), as obtained from the processed strong motion records.

In Table 1, Figures 5, 6 and 7 spectral amplification factors are given, where the effect of the nonlinearity is characterized by the coefficient c. This c coefficient is the ratio of SAF for the May 30 (or 31) 1990 Vrancea earthquakes to SAF for each stronger earthquake. S_a^* (g) and a^*(g) are the maximum spectral acceleration and, respectively, maximum acceleration if the system would have a linear response (behavior) to the fundamental period. For the Vrancea earthquake of May 31, 1990 (M_s=6.1) the response can be assumed to be in the elastic range, so that we have the possibility to compare the nonlinear effects with those predicted by a linear model. The present analysis indicates that the effect of nonlinearity could be very important. For instance, if we maintain the same amplification factor (SAF=5.8942) as for the relatively strong earthquake of May 31, 1990 with magnitude M_s =6.1, then at Bacau Seismic Station for the May 30, 1990 event (M_s =6.7) the peak acceleration has to be a^*_{max} =0.154g (+14.16%), while the recorded value was only, a_{max} =0.135g. Also, for the August 30, 1986 earthquake, the peak acceleration has to be a^*_{max}=0.107g (+45.57%), instead of the actual value of 0.0736 g recordcd at Bacau Seismic Station (Table 1).

The spectral amplification factors and, in fact, the nonlinearity, are functions of earthquake magnitude. In Table 2 we can see a strong nonlinear dependence of the spectral amplification factors on earthquake magnitude.

The amplification factors decrease as the magnitude increases. This is consistent with the data in Tables 1 and 2, which confirm that the ground accelerations tend to decrease as earthquake magnitude increases. As the excitation level increases, the response spectrum is larger for the linear case in comparison with the nonlinear one. This is consistent with the one-degree-of-freedom oscillator theory (used here to introduce the concept of spectral amplification factor), since the peaks of the displacement seismograms in the linear and nonlinear cases are controled by frequencies that are deamplified due to the nonlinearity.

The parameters of the seismic motion may considerably increase (with a percentage of 30%) if the wave front is characterized by higher frequencies and longer travel times, as a consequence of the altered structure (large deformations) of earth or superpositions of radiated waves from multiple sources. As a matter of fact, in the March 4, 1977 earthquake of Vrancea ($M_s=7.2$), for propagation of succesive shocks along a quite vaste fault such sequences were with different orientations and amplitudes during short time intervals. Usually, the standard design procedures disregard such instances, or special peak effects.

6. Conclusions

In conclusion, we may say that, despite the fact that seismic waves may not be characterized entirely by critical values in the range of the nonlinear effects, the special features of the major earthquakes in the Romanian Plain may induce various amplifications due to nonlinear effects. One of them is, for instance, the directivity effect, well documented in Vrancea earthquakes. Another is the decrease in peak variability among various sites. In addition, we stress upon the fact that since no third-order nonlinear effects were considered in the present analysis, the associated seismic effects may be enhanced on increasing the order of nonlinear contributions. This aspect will be subsequently analysed, together with refraction and reflection effects.

Acknowledgements

This work has been carried out within the framework of the „*Strong Earthquakes: A Challenge for Geosciences and Civil Engineering*" cooperation between the Romanian Group for Strong Vrancea Earthquakes in the Institute of Earth Physics, Magurele-Bucharest, and the Research Center 461 in Karlsruhe, Germany.

References

Aki, K., Local site effects on weak and strong ground motion, Tectonophysics, 218, 93-111, 1993;

Aki, K., Richards, P., Quantitative Seismology.Theory and Methods, W. H. Freeman and Company, San Francisco,USA, 1980;

Archuleta, R. J., Direct observation of non-linearity in accelerograms, in Proceeedings of the 2nd International Symposium On the Effects of Surface Geology on Seismic Motion, ESG, December 1-3, 1998,Yokohama, Japan, pp787-792.

Beresnev, I. A., Kuo-Liang Wen, Yeong Tein Yeh, Nonlinear soil amplification: its corroboration in Taiwan, Bulletin of the Seismological Society of America,Vol.85, No.2, 496-515, 1995;

Beresnev, I. A., Kuo-Liang Wen, Nonlinear soil response –A reality?, Bulletin of the Seismological Society of America,Vol.86, 6, 1964-1978, 1996;

Cornea, I., Misicu, M., Cojocaru, E., Vaicum Al., Steflea, V., Directivity effects of elastic wave reactions induced by multiple shocks. Application to the earthquake of March 4,1977, Rev. Roum. Sci.Tech.-Mec.Appl., Tome 25, 3, 341-351, Bucharest, 1980;

Cornea, I., Misicu, M., Wave directivity of multiple dislocations mechanisms with application to the analysis of kinemetic effects during the 4 March earthquake, Rev. Roum. Sci.Tech.-Mec.Appl., Tome 26, 5, 711- 723, 1981;

Doyle, T., Ericson J., Nonlinear theory of elasticity, Advances in Applied Mechanics, Vol.IV, Editors H.L. Dryden, Th.Von Karman, War-Marine Laboratory, Washington,(1956) transl. in Problemi Mehaniki, 11, 63-115, Moscow, 1959;

Field, E. H. et al. (11 co-authors), Nonlinear site response: where we are at? Seism. Res. Lett. Vol 69, 2, pp.230-234, 1998;

Hays, W.W., Procedings of Conference XXII, A Workshop on "Site-Specific Effects of Soil and Rock on Ground Motion and the Implications for Earthquake-Resistant Design", U.S.Geological Survey Open File Report 83-845;

Marmureanu, Gh. Cojocaru, E., Moldoveanu, C., The strong Vrancea earthquakes and the dynamic (spectral) amplification factors, Rev. Roum. Sci.Tech.-Mec.Appl., Tome 40, 2-3, 293-315, 1995;

Marmureanu, Gh., Moldoveanu C., Cioflan, C., The dependence of the spectral amplification factors of Vrancea earthquakes magnitudes, Rev. Roum. Sci.Tech.-Mec.Appl., Tome 41, **5-6**, 487-491, 1996;

Marmureanu, Gh., Moldoveanu, C., Cioflan, C., Apostol, B-F., The seismic earth response by considering nonlinear behavior of the soils at strong Vrancea earthquakes, Vrancea Earthquakes Monography: Tectonics, Hazard and Risk Mitigation, Kluwer Academic Publishers, Netherlands, p.175-185, 1999;

Misicu, M., Plane waves in nonlinear elastic layered halfspace, Rev. Roum. Sci.Tech.-Mec.Appl., in press;

Misicu, M., Marmureanu, Gh., Cioflan C., Apostol B-F., On nonlinear effects in waves propagation at the Earth surface and consequences for the seismic risk in the Romanian Plain, Rev. Roum. Sci.Tech.-Mec.Appl., in press;

Ni, Shean-Der, Siddhartahan, Raj V., Anderson, John G., Characteristics of non-linear response of deep saturated soil deposits, Bulletin of the Seismological Society of America, vol 97, no2, pp.342-355, 1997;

Pavlenko, O., Nonlinear seismic effects in soils: numerical simulation and study, Bulletin of the Seismological Society of America, vol 91, no2, pp.381-386, 2001;

Reiter, L, Earthquake Hazard Analysis .Issues and Insights, Columbia University Press, New York Chichester Chichester,West Sussex., 1991.

Signorini, A., Transformationi termoelastiche finite,caracteristiche dei sistemi dif-ferenziali, one di discontinuita,in particolare onde d'urto e teoria degli eplosi-vi, Atti 24th Reiunion Soc. Ital. Prozi.Sci., **3**, 6-26, 1936;

Yokohama Symposium, Proceeedings of the 2nd International Symposium "On the Effects of Surface Geology on Seismic Motion", ESG, December 1-3, 1998,Yokohama, Japan, 1998;

Yu, G., Anderson, J. G., Siddharthan, R., On the characteristics of nonlinear soil response, Bulletin of the Seismological Society of America, 83, **1**, 218-244, 1992.

Small-Scale Mantle Plumes: Imaging and Geodynamic Aspects

Joachim R R Ritter

Geophysikalisches Institut, Universität Karlsruhe, Hertzstr. 16, 76187 Karlsruhe, Germany, joachim.ritter@gpi.uni-karlsruhe.de

Abstract

Columnar, upwelling masses of hot rock in the Earth's mantle, called mantle plumes, are spectacular features of our planet because they supply the melts for the major intraplate volcanic sites. Although geodynamic modelling provides some constraints on the geometrical, physical, and fluid-dynamic properties of mantle plumes, there are still many uncertainties concerning the nature of mantle plumes, as it is difficult to image them inside the real Earth. These difficulties arise because mantle plumes are at depths of several tens to hundreds of kilometres, because they are relatively narrow (100–200 km), and because they have a small seismic velocity contrast (< 5%) relative to the surrounding mantle. Only with the help of specifically designed seismological experiments it was possible to image and identify a few mantle plumes in the last 10 years (e.g. Iceland, Massif Central, Eifel). Despite these advances a lot of uncertainties and unsolved questions remain about the physical properties of mantle plumes.

Current models for mantle plumes indicate that the buoyancy flux or mass deficit of the upwelling material varies by almost three orders of magnitude (~0.01-10 Mg/s). Besides the massive upward flows underneath large igneous provinces and major hotspots, there is now also evidence for small-scale mantle plumes. The later are related with very low or even no volcanic activity. Here I derive the physical properties (size, excess temperature, buoyancy flux, and heat flux) of the small-scale Eifel plume (Germany) based on seismological models. The Eifel plume has a width of about 100 km and extends from the upper asthenosphere (~70-80 km depth) down to at least the transition zone. Its excess temperature ΔT reaches about 100-150 K and the estimated buoyancy flux B is about 0.05 ± 0.04 Mg/s. For the neighbouring small-scale mantle plume underneath the French Massif Central a B of about 0.09-0.7 Mg/s is determined based on existing integrated seismological and petrophysical models ($\Delta T\sim150$-200 K, $r\sim60$-75 km).

Small-scale mantle plumes may contribute significantly to the mass and heat flow in the Earth's mantle, if they exist in a great number (e.g. more than 5000), as recently proposed by Malamud & Turcotte (1999). A major challenge for geoscientists in the next decades is to identify more of these small-scale plumes and provide unique models that are based on observations. Complementary fluid- and geodynamic simulations and petrophysical modelling is necessary to fully understand the dynamics of small-scale mantle plumes as well as their contribution to the mixing of mantle material.

1. Introduction

During the last decades of the 20th century the concept of mantle plumes was introduced and continuously improved to explain some aspects of Earth's mantle circulation and intraplate hotspots as well as volcanism on other planets (e.g. Mars). For a summary on plumes and mantle dynamics see e.g. Davies (1999) or Jackson (1998). Here I review the hypothesis that there must be mantle plumes of different scales or buoyancy fluxes, including small-scale plumes which are not explained by the classic models. Such small-scale mantle plumes are hardly explored up to now. A recent seismic study in the Eifel region, Central Europe, revealed a whole upper mantle plume with a diameter of about 100 km which is related to a tiny volume of surface volcanism (<20 km^3 volcanic rock). The seismological models are used to obtain further physical parameters such as excess temperature (the temperature difference between the plume and the surrounding 'normal' mantle) and buoyancy flux. This allows a comparison with numerical studies and other known mantle plumes. The results imply that more studies on plumes are required in the future to better understand their contribution to Earth's mantle dynamics.

2. Mantle Plumes

Centres of long-lived, relatively stationary and massive volcanism were named hotspots in the 1960s (Wilson 1963), when the concept of global plate tectonics with mobile lithospheric plates was developed. In 1971 Morgan argued that the deep-seated sources of the hotspots are hot flows of buoyant mantle material. Morgan (1971) called these upwellings (mantle) plumes (Fig. 1). Since then numerous field, laboratory and numerical studies were conducted to achieve a better understanding of mantle plumes and their relations to mantle mixing, global heat transfer, and surface volcanism. Today the properties of mantle plumes are defined mainly from results of geodynamic modelling and only rarely from observations in nature. Tank-experiments (e.g. Griffith and Campell 1990) and numerical modelling (e.g. Albers and Christensen 1996; Steinberger and O'Connell 1998) determined most characteristics of mantle plumes: They ascend from a thermal boundary layer (Fig. 1) as soon as the density and viscosity contrasts are large enough that the heated material has a positive buoyancy. Due to the resistance of the overlaying mantle, the rising diapir develops a more or less broad plume head, depending on the viscosity contrast. The upwelling material is concentrated inside a channel (plume conduit or plume stem) which may be tilted by the surrounding convecting mantle (mantle wind) (Steinberger and O'Connell 1998). If the rising plume hits an obstacle or if the buoyancy diminishes, the hot material spreads laterally and forms a broad mushroom-like plume head (Fig. 1). Decompressional melting is controlled by the excess temperature and the pressure release in the shallow asthenosphere (< ~50-100 km depth). The amount of parental melt is only a few percent compared to the total plume material (e.g. Davies 1999; Ribe and Christensen 1999). The pressure release and hence the amount of melt production are more effective underneath thin oceanic lithosphere compared to thick continental lithosphere (White 1993). Depending on the location of the thermal boundary layer at depth (e.g. the core-mantle boundary or the mantle transition zone), mantle plumes transport mass and heat through various parts of the mantle and thus they contribute to mantle mixing. There is also a debate whether whole mantle plumes are a principle cooling mechanism for the heat that leaves the core (~5.7 TW, Anderson 1999).

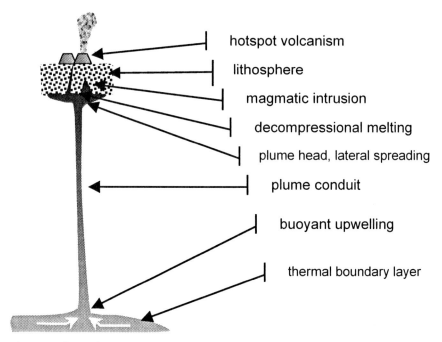

Fig. 1. Schematic sketch of a buoyant mantle plume that is upwelling from a thermal boundary layer

There are no strict definitions to classify plumes concerning their 'size' and obviously mantle plumes cannot be observed 'directly'. Only the surface volcanism and the erupted material is accessible, but these masses are not an appropriate parameter to describe the plume size as decompressional melting depends on the thickness of the overlaying plate (White 1993) and as the ascent of the melts depends on the connectivity of the partial melt batches as well as on (pre-)existing pathways through the overlying lithosphere. Thus plumes with similar physical properties in the mantle may erupt quite different amounts of magma. A better way to characterise the plume size is the mass flow deficit or the buoyancy flow B in the mantle. Following Sleep (1990) the buoyancy flow B of a rising plume (column/blob) can be estimated by the density contrast $\Delta\rho$ of the rising material and volume flux Q_p:

$$B = \Delta\rho_m \, Q_p = \rho_m \, \alpha \, \Delta T \, Q_P \qquad (1)$$

with ρ_m (mean) mantle density, α thermal expansion coefficient of the mantle material, and ΔT excess temperature of the plume. The volume flux is a function of the rise velocity v_{pl} which can be estimated e.g. from

Poiseuille's equation for flow inside a tube adopted to a buoyant viscous fluid inside the Earth (Christensen, pers. comm.):

$$v_{pl} = \frac{3 \, \rho_m \, \alpha \, \Delta T \, g \, r^2}{8 \, \eta} \tag{2}$$

r is the radius of the tube (plume conduit), g is Earth's gravity acceleration, and η is the viscosity of the flowing material. A factor 3/4 is included because it is assumed that the temperature anomaly inside the plume conduit decreases from the maximum excess temperature in the centre (between $r=0$ and $r=1/2 \, r_{max}$) at $1/2 \, r_{max}$ to the ambient mantle temperature at radius r_{max}. Thus a volume flux of

$$Q_P = \frac{1}{4} \, \pi \, r^2 \, v_{pl} \tag{3}$$

results in which $r \, (= r_{max})$ is the radius of the plume conduit. Combining Eqs. (1) to (3) gives

$$B = \frac{9 \, \pi \, g}{64} \frac{\left(\rho_m \, \alpha \, \Delta T \right)^2 \, r^4}{\eta} \tag{4}$$

Thus the buoyancy flux depends on relatively constant parameters in the mantle such as g, ρ_m and α as well as 'plume'-dependent variable parameters such as ΔT, r and η. $B(r,\Delta T)$ is highly non-linear because it depends on the radius in fourth order and the excess temperature in second order. Additionally, the excess temperature controls the viscosity of the plume material via an Arrhenius law. Mantle viscosity depends on the strain rate ($\dot{\varepsilon} = \partial \varepsilon / \partial t$) and the shear stress σ at depth:

$$\eta = \frac{1}{2} \frac{\sigma}{\dot{\varepsilon}} \tag{5}$$

Following Karato and Wu (1993) the strain rate in the mantle can be approximated by the strain rate of olivine which is the main constituent mineral of mantle material.

$$\dot{\varepsilon} = A \left(\frac{\sigma}{\mu} \right)^n \left(\frac{b}{d} \right)^m e^{-\frac{E+PV}{RT}} \tag{6}$$

The strain rate depends on a pre-exponential factor (A), the shear modulus (μ), the Burger's vector length (b), the grain size (d), two exponential factors n and m, the activation energy E, the activation volume V, the pressure P and the Gas constant R. In the following course, I use strain rates that are based on diffusion creep at depths below 100 km together with constant values for b, d, A, E, V, n and m as adopted from Karato and Wu (1993) and listed in Table 1. P is determined by the lithostatic overburden. The resulting average mantle viscosity at a potential temperature of 1300 °C is about $5 \cdot 10^{20}$ Pa s and it is in agreement with other studies (e.g. Kaufmann and Wu 2002).

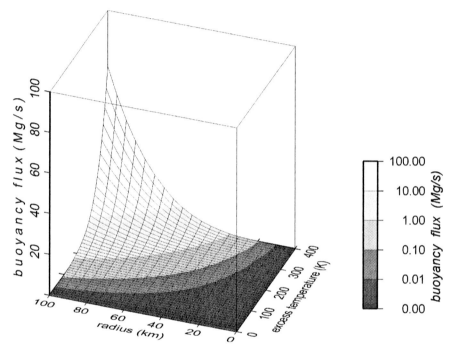

Fig. 2. Hypothetical range of buoyancy fluxes of mantle plumes as function of conduit radius and excess temperature

In Fig. 2 the buoyancy flux B through plume conduits is given in Mg/s as a function of plume radius r and excess temperature ΔT in the upper mantle. A potential background temperature of 1300 °C is assumed (McKenzie and Bickle 1988). The maximum r of 100 km and the maximum ΔT of 400 K is taken from several geodynamic studies. The non-linearity of $B(r,\Delta T)$ is obvious in Fig. 2. Below r~25 km only for large ΔT (~350 K) a noticeable buoyancy flux is present. For small ΔT (~70 K) a noticeable buoyancy flux needs a radius of about 75 km. To get a B of at least 1 Mg/s it takes r and ΔT of 100 km and 180 K or 50 km and 290 K, respectively. For B~10 Mg/s it takes r and ΔT of 100 km and 280 K or 60 km and 400 K. With the maximum values of r=100 km and ΔT=400 K, a large B of 72 Mg/s is found.

Published estimates of B for plumes in the Earth range from 0.3 Mg/s (Bowie plume) to 8.7 Mg/s (Hawaii) (Davies 1988; Sleep 1990). Detailed numerical studies for Hawaii and its swell resulted in 2.2 Mg/s<B<3.5 Mg/s (Ribe and Christensen 1999) and B~3.98 Mg/s (Zhong and Watts 2002). However, using Zhong and Watts (2002) estimate for r (50-70 km) and ΔT (400 K), I arrive at B~4.5-17.3 Mg/s with Eq. (4) and the parameters in Table 1. In Fig. 3 the cumulative number of 43 plumes (open circles) is plotted as function of B. Large error bars (at least ±50 %) must be assigned to the displayed values because the input parameters are often poorly constraint. However, it seems that there is a power law distribution between the frequency and size of plumes that implies a huge number of small plumes which have not been discovered yet (Malamud and Turcotte 1999).

Table 1. Physical parameters of the upper mantle used for this study

parameter	symbol	value	SI unit
[a] pre-exponential factor (viscosity law)	A	$7 \cdot 10^{15}$	-
[a] exponent viscosity law	m	2.5	-
[a] exponent viscosity law	n	1.0	-
[a] activation energy	E	$2.7 \cdot 10^5$	J mol^{-1}
[a] activation volume	V	$5.5 \cdot 10^{-6}$	m^3 mol^{-1}
[d] gas constant	R	8.3143	J K^{-1} mol^{-1}
[b] Burger's vector	b	$0.5 \cdot 10^{-9}$	m
[b] grain size	d	$1 \cdot 10^{-3}$	m
[b] shear modulus	μ	$8 \cdot 10^{10}$	Pa
[c] potential background temperature	T	1300	°C
[d] mean mantle density	ρ	3400	kg m^{-3}
[d] thermal expansion coefficient	α	$4 \cdot 10^{-5}$	-
[d] specific heat capacity	C_p	1200	J kg^{-1} K^{-1}
[d] gravity acceleration	g	9.81	m s^{-2}

[a] mean value between dry and wet olivine after Karato and Wu (1993)
[b] after Karato and Wu (1993)
[c] after McKenzie and Bickle (1988)
[d] after Ahrens (1995)

Numerical modelling results indicate that plumes which start at the core-mantle boundary need at least B>1 Mg/s to remain hot enough that melting can occur below the lithosphere (Albers and Christensen 1996). Plumes with smaller B and magma production should either start from the transition zone or their melting should involve volatiles which reduce the solidus (Albers and Christensen 1996). In the following text a B<1 Mg/s criterion is defined to characterise small-scale mantle plumes. Using Eq. (4) and

Fig. 2 this corresponds to upper mantle structures with approximately 0T<80 K and r< 100 km as well as seismic velocity contrasts of less than 3%. These relatively small values and additional effects such as wavefront healing and noisy data make obviously seismic imaging of small-scale plumes a difficult task.

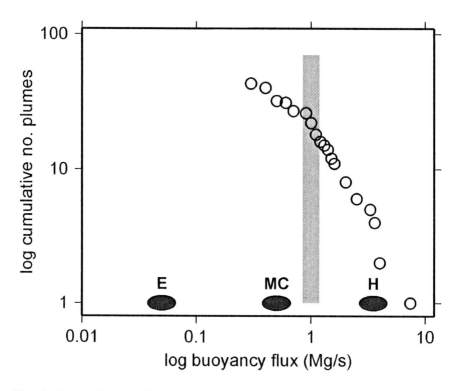

Fig. 3. Cumulative number of mantle plumes as function of buoyancy flux *B* (open circles, modified after Malamud and Turcotte 1999). The solid ellipses are estimates for *B* of the Eifel and Massif Central plumes (this study) and Hawaii (Ribe and Christensen 1999). The vertical bar at *B*~1 Mg/s indicates the border between small-scale plumes and bigger classic plumes

3. Observations

At the Earth's surface only the consequences of plumes such as hotspot volcanism, heat flow, uplift, or gravity anomalies can be observed directly. For example voluminous lava effusions in large igneous provinces are attributed to enormous melting events in huge plume heads (Eldholm and Coffin 2000). The isotopic and geochemical compositions of hotspot lavas are used to derive the deep origin of the plume material. However, these models are based on assumptions about the global distribution of the analysed elements, isotopes and minerals in the Earth. With the exception of seismology, geophysical techniques do not have the required resolution (< 100 km) to detect plume-like structures in the lower part of the upper mantle or even inside and below the transition zone. A summary of tomographic studies of mantle plumes is given in Nataf (2000). Even with the most modern seismological methods it is not easy to image unambiguously a mantle plume. Global tomography models, based on millions of travel times from seismic body-waves, have a maximum resolution of about 100 – 500 km in the better resolved regions. Plumes detected by global tomographic studies (e.g. Zhao 2001) are much wider (300-800 km) than expected from geodynamic modelling (100-200 km). This is mainly due to the sparse seismic ray distribution that requires still a coarse parameterisation (100-500 km) in global tomography. Furthermore, spatial smearing of the velocity perturbations due uneven ray distribution causes imperfect resolution of structures.

Seismological experiments with well designed station configurations above a suspected plume are more promising. A dense station spacing (< 20 km) allows to map seismic velocity perturbations of a few percent with a size of a few tens of kilometres in the upper mantle. However, mobile station deployments have a restricted aperture of maximum a few hundred kilometres due to the limited number of available instruments. This reduces resolution at great depth due to the lack of crossing rays in the tomographic model space. Anyway, plume-like low-velocity structures in the upper mantle were found with such experiments underneath Iceland (Wolfe et al. 1997; Foulger et al. 2000) and the French Massif Central (Granet et al. 1995a,b).

An alternative technique which gained important results in the last few years is the receiver function approach which uses teleseismic P-SV converted phases. Underneath Hawaii (Li et al. 2000), Iceland (Shen et al. 1998 and 2002) and Eifel (Grunewald et al. 2001) there is evidence for depth variations of the seismic discontinuities at the top and bottom of the transition zone. These deflections are attributed to the excess temperature of the plumes, because the additional heat changes the transformation depth (pressure) of the mineral phases which cause the discontinuities.

The limited number of high-resolution data sets is presumably the main reason why only a few mantle plumes have been discovered at depth yet. There are also some discrepancies between plume structures obtained from geodynamic modelling and the seismic images. The proposed plume heads have not been detected in seismic images up to now. This may be due to a lack of resolution in the tomographic images or a too large viscosity contrast in the modelling attempts.

4. Eifel Plume

4.1 Seismological Anatomy

Recent results obtained by the Eifel plume project represent a successful attempt to image a small-scale mantle plume underneath Central Europe. Previous geophysical, geological, geochemical and geodynamic studies had collected evidence for a possible plume-like anomaly underneath the Quaternary Eifel volcanic fields (for a summary see Fuchs et al. 1983).

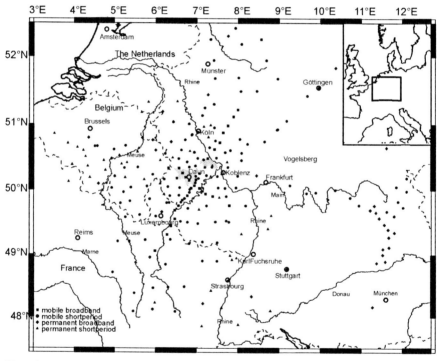

Fig. 4. Station distribution of the Eifel plume project. 158 mobile instruments (squares and dots) were deployed between 84 permanent seismological observatories (diamonds and triangles). The two Quaternary Eifel volcanic fields are outlined in grey. The inset shows the geographical position of the study area within Central Europe

Following earlier Mesozoic and Tertiary volcanism, about 600-700 ka ago two new volcanic fields started their activity in the West and East Eifel (Fig. 4). Over the same time period up to 250 m of uplift occurred (Meyer and Stets 1998).

In 1997 the European Eifel plume project was initiated to get a much clearer image of the crust and upper mantle under the Eifel region (Ritter et al. 1998a). The seismological group with scientists from Belgian, Dutch, French, German and Luxembourg institutions operated a temporal station network that was carefully designed to record the necessary data (Ritter et al. 2000). Altogether 158 mobile and 84 permanent stations (Fig. 4) were used to observe local and global seismic events during the field experiment from November 1997 until June 1998.

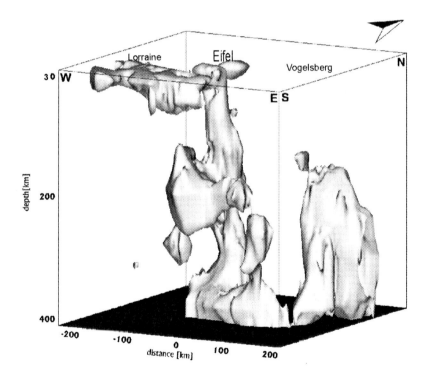

Fig. 5. 3D representation of the P-wave low-velocity anomaly underneath the Eifel region. The 400 km by 400 km wide model represents the -1% v_p anomaly in the upper mantle (centre at 7.33 °E and 49.75 °N)

To determine the 3D seismic velocity structure down to the top of the transition zone, travel time residuals from 7319 teleseismic P-wave first arrivals were analysed (Ritter et al. 2000,2001). In Figure 5 the –1% isovelocity anomaly relative to the IASP91 reference Earth model (Kennett and Engdahl 1991) is displayed as determined by nonlinear inversion (Ritter et al. 2001). The vertical extension of the model corresponds to the whole upper mantle from 30 to 410 km depth and the centre is at 7.33 °E and 49.75 °N. Crustal structures are not well resolved due to the noncrossing subvertically incoming rays. There are three low-velocity bodies in Fig. 5. The main inversion result is a columnar low v_p anomaly with a lateral contrast of up to -2%. This ~100 km wide structure extends to at least 410 km depth and it is situated below the West Eifel volcanic field (Fig. 4). To the south-west a shallow (<50 km depth) low v_p anomaly underneath the French Lorraine region is found which is possibly an inversion artefact caused by downward smearing of an unknown crustal structure. In the eastern part of the model, below ~200 km depth, another low-velocity body can be seen (Fig. 5). This anomaly is situated below the Vogelsberg area which was a major volcanic site in Miocene time. However, due to low resolution the reliance to the Vogelsberg anomaly is quite low because it is traversed by only few rays and large blocks are used for the parameterisation (Fig. 6). This causes an overestimation of the size of structural elements.

Several resolution analyses confirmed that the main low-velocity anomaly underneath the Eifel region reaches down to the top of the transition zone, although the amplitude of the velocity perturbations is underestimated (e.g. Ritter et al. 2001). In Fig. 6 the results of a reconstruction test for the P-wave model is displayed at 8 depth intervals. As input anomalies two checkerboard structures were placed into the model space with ±2% variation in P-wave velocity relative to the background velocity. This resolution test can be used to determine the horizontal resolution of structures and the vertical smearing along the teleseismic raypaths above and below the two checkerboard anomalies. Synthetic travel times were calculated with a 3D raytracing routine. The ray geometry of the 7319 rays from the actual tomography experiment is used. Additionally, Gaussian noise is added with an amplitude as expected for the real data. The parameterisation is the same as for the actual tomography in Fig. 5 and it is chosen according to the ray density. The first checkerboard was inserted at 31-70 km depth (Fig. 6b) with a lateral extent of the anomalies of 39 km by 39 km. The second checkerboard is placed at 170-220 km depth (Fig. 6e) with a lateral extent of the anomalies of 50 km by 50 km.

Directly underneath the dense station network (at ± 200 km distance from the centre at 7.33 °E and 49.75°N) the horizontal resolution of the 39^2

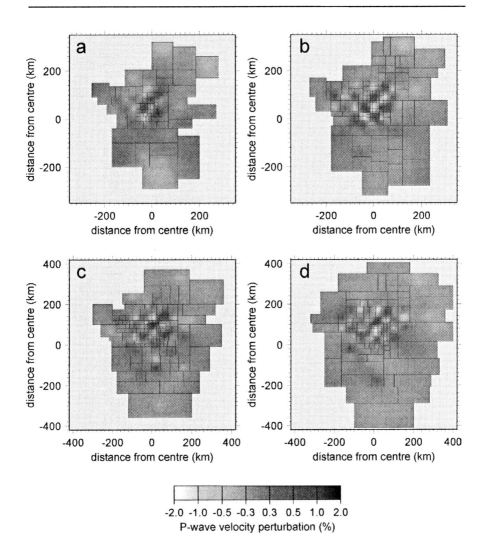

Fig. 6. Results of a reconstruction test for the model in Fig. 5 to determine the resolution properties of the inversion. As input checkerboard anomalies were placed at 31-70 km (b) and 170-220 km (e) depth with ±2% variation in P-velocity relative to the background velocity (IASP91). The displayed model layers are at the following depths (in km): (a) 5-31, (b) 31-70, (c) 70-120, (d) 120-170, (e) 170-220, (f) 220-270, (g) 270-340, and (h) 340-410. The parameterisation of the inversion model is indicated by boxes and it is the same as for the actual tomography in Fig. 5.

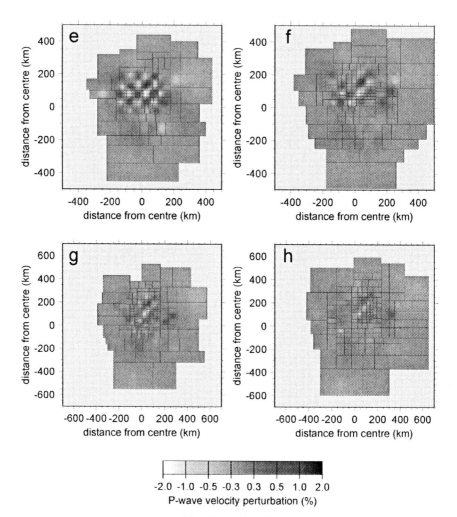

Fig. 6. continued

and 50^2 km^2 wide structures is quite good (Fig. 6b&e). Towards the edges of the model space the resolution is low due to missing criss-crossing rays and the coarse parameterisation. Vertical smearing causes blurring of the checkerboard structures to the next upper and lower layer, especially near the surface (Fig. 6a). This also causes a reduction of the input anomalies from 2% to about 1%. The input structure at 170-220 km depth is well recovered but smearing is observed to the next deeper layer (Fig. 6f). Below, at 270-410 km depth the smearing is nearly absent (Fig. 6g-h). This implies that deep anomalies in the real model (Fig. 5) are not due to downward projection of a shallow low-velocity body by the inversion process. Further tests demonstrate that anomalies below 200 km depth are systematically underestimated in amplitude by as much as 75%. This finding means that the actual anomaly of the Eifel plume is more than the 1% v_p perturbation as it is found in our inversion results.

The inversion of 3773 travel times of first arrivals from teleseismic S-waves shows also a columnar low-velocity structure. The v_s contrasts vary between –5% below the Moho and –1% in the lower asthenosphere down to more than 400 km depth (Keyser et al. 2002). The v_s anomaly is wider than the v_p anomaly. This could be due to the much larger wavelength of the S-waves (up to 60 km) compared to the P-waves (~10 km). Again resolution tests demonstrate that the amplitudes of the anomalies are significantly underestimated below 200 km depth. An interesting feature in the S-wave model is a 'hole' of the low-velocity anomaly between about 180±20 km and 250±20 km depth. Here the negative v_s perturbation is nearly vanishing. A possible explanation for this 'hole' are dehydration processes (Keyser et al. 2002).

The attenuation properties of the lithosphere and asthenosphere in the Eifel region were investigated with teleseismic P-waves. Applying the teleseismic fluctuation wavefield method (Ritter et al. 1998b) to P-coda data reveals a significant increase of the scattering strength in the area of the Eifel volcanic fields relative to the surrounding regions (Rothert and Ritter 2000). The increase of the scattering strength is about 30% and corresponds to an increase in $\sigma^2 a$, the squared rms velocity fluctuations σ (in percent) and the average size a of the heterogeneities in the lithosphere. This scattering anomaly underneath the Eifel is interpreted as an intrusion zone (solidified dykes and magma chambers) related to the magmatic processes.

To determine quantitatively the attenuation of the teleseismic wavefronts in the Eifel region and the surrounding Rhenish Massif, the damping of the P-wave pulse was examined by determination of the t^* parameter in the frequency domain as well as by spectral ratios δt^*. t^* is proportional to the inverse of the quality factor Q and hence attenuation. The observed damping operators t^* have systematic variations at the station sites, indicating higher t^* values (~0.1 s) -and hence stronger damping- in the volcanic Eifel region (Meyer 2001). In the same area increased values of the spectral ratio δt^* relative to a reference station (BFO) outside the Rhenish Massif are found. After eliminating source-side attenuation effects, residual damping parameters δQ reveal a -200±50 Q anomaly under the Eifel region relative to the surrounding Rhenish Massif (Meyer 2001). For a tomographic inversion 3764 spectral ratios of teleseismic P-wave first arrivals were determined. A cluster of significantly increased δt^* values with up to +0.8 s clearly outlines a high-attenuation anomaly below the Cenozoic Eifel volcanic fields. In the inversion model three main anomalies are identified and interpreted: First, the strongest damping is observed in the lithosphere underneath and surrounding the East Eifel volcanic field. This is interpreted as scattering attenuation at a magmatic intrusion zone that was also identified by the P-coda study described above. Second, an asthenospheric anomaly (above ~200 km depth) is found underneath the West Eifel volcanic field. It coincides spatially with the seismic low-velocity region in the P- and S-wave travel time tomographies. Third, at about 250 km depth another absorption anomaly is detected which reaches down to more than 300 km depth. A less attenuating region at about 200 km depth coincides spatially with the 'hole' in the S-wave model (see above).

An independent study based on data of the German Regional Seismic Network (GRSN) and a GEOFON station (WLF in Luxembourg), using the receiver function methodology, identified an up to 20 km downwraping of the 410 km discontinuity underneath the Eifel region (Grunewald et al. 2001). This corresponds to a positive temperature anomaly of about 250 K (Bina and Helffrich 1994). In contrast the 660 km discontinuity seems to be flat at the expected reference depth (Grunewald et al. 2001), indicating that the Eifel plume may originate in the transition zone. Analysing additional data from the Eifel plume seismological network, Budweg et al. (2001) come to the same conclusion with a high-resolution receiver function study. Additionally, they find further anomalies in the mantle that are consistent with the tomographic models (Budweg, pers. comm.).

4.2 Excess Temperature and Buoyancy Flow

The modelled velocity perturbations can be used to determine the excess temperature of the Eifel plume. Seismic velocity is a function of composition (mineral phases, melts, fluids, ..), pressure and temperature. Here seismic anisotropy is neglected because it can be excluded as a first-order effect related to the seismic velocity reduction in the tomographic models (Keyser 2001; Keyser et al. 2002). As mineral composition in the shallow mantle a spinel lherzolite is assumed with an 'average Eifel' composition of 75% olivine (Ol), 17% orthopyroxen (OPx), 7% clinopyroxene (CPx) and 1% spinel (Sp) (Witt-Eickschen, pers. comm.). Petrophysical relations of mineral parameters, including the anharmonic temperature $(\partial v / \partial T)$ and pressure $(\partial v / \partial P)$ derivatives for seismic velocities as well as the anelastic contributions (Karato 1993), are taken from the literature. Calculations for the composite mantle rocks are based on the Voigt-Reuss-Hill averaging scheme. At ambient P-T-conditions in the uppermost mantle a 1% v_p or v_s reduction corresponds to about 90 K or 60 K increase in temperature, respectively. This is in agreement with laboratory measurements at olivine polycrystals (T=1200 °C and Q~100) at solid-state conditions (Jackson et al. 2001). A linear extrapolation of these relations results in up to 150-300 K excess temperature for the Eifel plume (-2% v_p and –5% v_s). However, the velocity-temperature relationship becomes non-linear at high temperature. Then the temperature sensitivity of v_s increases, and thus smaller temperature variations cause the velocity contrasts (Jackson et al. 2001). This effect may lower the above estimated excess temperature to about 150-200 K. Using experimental relationships by Kampfmann & Berkhemer (1985) or Sato et al. (1989) for peridotite, a temperature increase of 100-150 K can explain the observed δQ of about -200 by solid-state relaxation processes in the upper mantle (see also Tan et al. 2001).

An excess temperature of up to 200 K for the Eifel plume is reasonable because the solidus temperature must be crossed at least in some parts of the anomaly to allow the onset of magmatic processes. However, a thermal anomaly of maximum 200-300 K is much more than it is necessary to produce the small amount of erupted material in the Eifel (< 20 km^3). Therefore, additional effects should be considered that might also reduce the seismic velocity. The occurrence of partial melt in the lower lithosphere and upper asthenosphere is justified by the basaltic Quaternary volcanism. Small amounts of partial melt, distributed in penny-shaped melt inclusions in the rock fabric, reduce v_p and v_s by up to 1.8% and 3.3% per 1% melt (Faul et al. 1994). Thus the combined effect of about 1% partial melt plus 100-150 K excess temperature can explain the observed v_p and v_s contrasts. Regions without melt should be ~150 K hotter than the surrounding mantle.

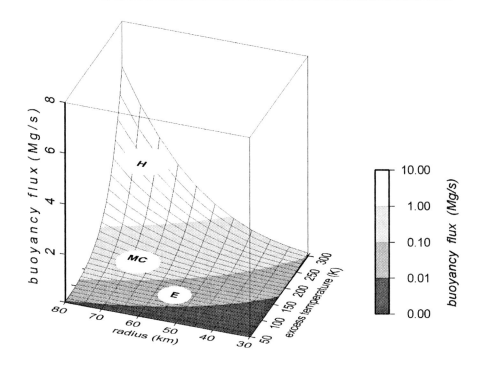

Fig. 7. Plausible buoyancy fluxes of mantle plumes as function of radius and excess temperature. E: Eifel, MC: Massif Central, H: Hawaii

To determine the excess temperature in the lower asthenosphere at about 300 km depth an anhydrous garnet peridotite (60% Ol, 25% CPx and 15% Garnet (Gt)) is assumed. The anharmonic and anelastic temperature-velocity relations at ambient P-T-conditions predict that a 1% v_p or v_s reduction is caused by ~70 K increase in temperature. This implies an excess temperature of the Eifel plume of at least 70 K at this depth but it is most likely larger, because the amplitude of the velocity perturbations (-1% v_p and –1% v_s) below 200 km depth are underestimated due to smearing in the tomographic modelling, as it is found with reconstruction tests (see above and Ritter et al. 2001; Keyser et al. 2002). Using a Clausius-Clapreyron slope of 2.9 MPa/K (Bina and Helffrich 1994), the 20 km downwraping of the 410 km discontinuity as determined by receiver function migration (Grunewald et al. 2001) is compatible with a positive temperature anomaly of about 250 K. This large excess temperature also indicates that the velocity contrasts in the tomographic models are underestimated.

For the next considerations I use the following seismologically constrained parameters for the Eifel plume: radius r = 50-60 km, excess temperature $\Box T$ = 100-150 K and upper mantle parameters as given in Table 1. Note that the uncertainties related with the parameters in Table 1 are much smaller than the uncertainties in the seismological model. Using the above values, the Eifel plume rise velocity v_{pl} (Eq. 2) ranges from 0.015-0.06 m/a (or 1.5-6 km per ka). The corresponding volume flux Q_p (Eq. 3) is 0.03-0.18 km^3/a which leads to an estimated buoyancy flux B (Eq. 4) of 0.01-0.09 Mg/s for the Eifel plume.

Assuming as much as 1% partial melting of the plume material, as estimated from the seismic velocity perturbations, we arrive at a melting rate of 0.3-1.8 10^{-3} km^3/a. This value can be compared to the observationally determined melting rate in the uppermost mantle of approximately 0.1-0.15·10^{-3} km^3/a beneath the Eifel volcanic fields (Wörner 1998). The later is calculated from the observed cumulative eruption rate at the surface over the life time of the Quaternary volcanism of about 600 ka and relationships between evolved and parent primitive magmas. A comparison between the seismologically estimated and observationally determined melting rates implies that the above derived buoyancy flow range seems to be more realistic at its lower values. However, upper mantle diapirism due to very small buoyancy flows of $B \sim 0.01$ Mg/s is unrealistic, because the heat dissipation would stop the driving mechanism (Albers and Christensen 1996). A minimum $B \sim 0.05$ Mg/s seems to be required for small-scale plumes that rise from the transition zone. Thus the average determined value of $B \sim 0.05$ Mg/s is more realistic based in geodynamic simulations, although it overestimates the observed melting rate by a factor of up to 10. This discrepancy can be reduced by either assuming a lower portion of partial melt or by a more effective buoyancy mechanism for small-scale temperature anomalies, e.g. by including volatiles to reduce the viscosity.

Using the volume flux Q_p (Eq. 3) it is possible to estimate the related heat flux Q_H:

$$Q_H = Q_P \, \rho_m \, C_P \, \Delta T \qquad (7)$$

with C_P the specific heat capacity (I take the value of olivine at ambient mantle conditions: ~1200 J kg^{-1} K^{-1}). For the Eifel plume this results in Q_H=0.4-3.5 GW.

4.3 Comparison with Massif Central and Hawaii

The same estimation for B and Q_H can be done for the mantle plume underneath the French Massif Central. The seismic images by Granet et al. (1995a&b) give an average radius of about 60-75 km. Sobolev et al. (1997) calculated an excess temperature of about 150-200 K, including effects such as the presence of partial melts. This leads to B=0.09-0.7 Mg/s (Fig. 7) which is at least 10 times more than underneath the Eifel. This value is compatible with the erupted material. Assuming an eruption volume of about 5000-6000 km^3 (Wilson, pers. comm.) and about 5 times more melted material in the mantle gives about 25-30·10^3 km^3 parental magma generation during the last 20 Ma (main volcanic episode). These values lead to an estimated melting rate of 1.2-1.5·10^6 m^3/a, again roughly 10 times the value for the Eifel. Thus both, the melting rate and B, are roughly balanced compared to the Eifel. The estimated heat flow for the Massif Central is 3.5-27.5 GW.

Davies (1999) gives an eruption rate of about 30·10^6 m^3/a for Hawaii. Again, assuming 5 times more parental magma in the mantle leads to a melting rate of 150·10^6 m^3/a (1000 times the Eifel value). The discussed values for B of 2-6 Mg/s (about 100 times the Eifel value) are relatively small compared with the melting rate, but B is better constraint than the melting rate.

5. Consequences and Discussion

The occurrence of small-scale mantle plumes poses several important questions: What are the physical conditions for the ascent of small-scale plumes? Where do these plumes come from? How many such plumes are in the Earth's mantle? How do these plumes contribute to the redistribution of heat and mixing of material in the Earth's mantle?

5.1 Origin and Ascent

With respect to general mantle dynamics, the first question is important because numerical modelling of (dry) plume upwelling has shown that at least a B of 1 Mg/s is necessary for plumes that start at the core-mantle boundary and that are still hot enough to produce melts (Albers and Christensen 1996). If small-scale plumes start in the transition zone and additional heat is provided at the exothermic phase transition at 410 km depth, then their ascent through the upper mantle is possible. Small-scale plumes, rising from the transition zone, could represent upper mantle blobs derived from much bigger lower mantle plumes (Brunet and Yuen 2000). The deeper upwellings can be retained by the endothermic 660 km phase discontinuity (e.g. Marquart and Schmeling 2000; Brunet and Yuen 2000). Afterwards, batches of material may penetrate into the transition zone where they start to ascend rapidly. Such a mechanism is one scenario for the small-scale upper mantle plumes in Europe (Massif Central, Eifel, Bohemia?, see Granet et al. 1995b Fig. 8). Global tomography identified a broad, 1000 km by 1000 km wide low-velocity anomaly underneath Central Europe (Goes et al. 1999). This anomaly may be the deep source of the secondary small-scale plumes in the upper mantle. A similar scenario is possible for the African and Southern Pacific superplumes which are repeatedly identified in the lower mantle by global tomography (e.g. Zhao 2001) and which are overlain by several smaller hotspots.

To facilitate the upwelling of small volumes of plume material, the required reduction of the viscosity (Eq. 4) may be caused by water in addition to the excess temperature. Since the minerals in the transition zone (β- and γ-phases of $(Mg, Fe)_2 SiO_4$ in wadsleyite and ringwoodite) can solve a few percent of water (Kohlstedt et al. 1996), it should be present in the upwelling material. Laboratory studies suggest that a few percent water increases the creep rate, and hence decreases the viscosity (Eq. 5), by as much as one order in magnitude (e.g. Kohlstedt et al. 1996). It should be noted that water-dependent viscosity is not used in numerical studies. Thus, the ascent of a small-scale plume from the transition zone with a small fluid constituent is quite reasonable.

5.2 Scales of Plumes

The above discussion and Fig. 3 imply that there are different scales of plumes. Furthermore, there must be much more small-scale plumes than have been identified yet, if a power-law distribution is behind the scale-frequency relations as proposed by Malamud and Turcotte (1999). However, as outlined in section 2, it is not easy to identify a small-scale plume with geophysical methods. If such an upwelling reaches the base of thick continental lithosphere, it may be too weak to produce significant amounts of melt. Due to the depth of ~200 km decompressional melting is prohibited and no surface volcanism occurs. In contrary, if a small-scale plume impinges the base of thin oceanic lithosphere at 50-100 km depth, it can generate small amounts of melt which are able to penetrate the plate and build up an intraplate seamount. An estimate by Wessel and Lyons (1997) gives as much as 70.000 seamounts for the Pacific plate. If just every 1/70 seamount is related to a small-scale mantle plume, then there existed 1000 of them underneath the Pacific plate during the last 200 Ma.

In Fig. 8 plumes of different scales are displayed schematically. The classic plumes (Fig. 8a&f) are indicated as voluminous upwellings which penetrate the lithosphere and produce a classic hotspot and a swell (e.g.

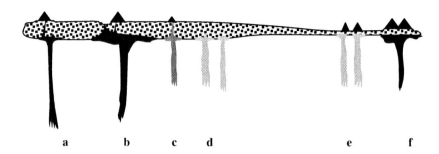

Fig. 8. Schematic sketch of plumes of different scales. a) Classic plume underneath a continent with major volcanism and a swell, e.g. Yellowstone. b) Massiv plume underneath a continent causing rifting, e.g. Kenya plume. c) Small-scale plume underneath a continent with minor volcanism, e.g. Eifel, Massif Central. d) Small-scale plume underneath a continent without volcanism. e) Small-scale plume underneath oceanic lithosphere generating a seamount. f) Classic plume underneath oceanic lithosphere generating an oceanic plateau or swell, e.g. Kerguelen plateau

Yellowstone in a continental environment or Reunion as an oceanic example). If large amounts of material are supplied and if the stress field is appropriate, then the continental plate fails and rifting is initiated (Fig. 8b). Classical examples are the Kenya and Ethiopia plumes in the East African rift system. Small-scale plumes produce only tiny amounts of melt and are related with minor (Fig. 8c) or no (Fig. 8d) surface volcanism due to the thick continental lithosphere which prevents decompressional melting. In contrast, the thinner oceanic lithosphere allows the production of melts and the generation of seamounts (Fig. 8e).

6. Future Perspectives

Small-scale mantle plumes are an attractive and important target for geoscientific research in the 21st century. As outlined above, imaging them is a major challenge as even large-scale plumes are rarely well resolved in seismological studies. Above all, the most important prerequisite is the availability of much more seismic instruments to achieve the necessary dense station coverage for high-resolution tomographic studies. Of special importance are ocean bottom instruments because mantle plumes under the oceans have hardly been studied in the past. Other important issues are high-pressure and high-temperature experiments with hydrous mantle rocks because their rheological behaviour is quite distinct from anhydrous rocks. The results of such laboratory studies are required to improve numerical plume simulations with hydrous rheologies. The petrophysical laboratory experiments are also needed for the interpretation of seismological models. Such an interdisciplinary approach will be the key for a better understanding of the quantitative contribution of the different types of plumes (outlined in Fig. 8) to the global heat flow, mantle circulation and the distribution of elements inside the Earth.

Acknowledgements

I thank Karl Fuchs for his continuous interest and support, especially for his encouragement for doing interdisciplinary work. U. Christensen provided the formula with Poiseuille's tube flow. M. Jordan calculated the resolution test in Figure 6. I. Jackson provided preprints. U. Achauer, U. Christensen and M. Martin provided useful reviews and comments. The Eifel plume team conducted the strenuous data collection for the tomography study. Mobile recording instruments used for the Eifel plume project were supplied by the GeoForschungsZentrum Potsdam, the French Lithoscope Pool, the Universities of Bochum, Karlsruhe and Potsdam, the Royal Observatory of Belgium and the Network of Autonomously Recording Seismographs (NARS) at Utrecht University. Further waveform data and instrument response functions were provided by Bundesanstalt für Geowissenschaften und Rohstoffe (BRG) Hannover, Germany (GRSN & GRF); GeoForschungsZentrum Potsdam, (GEOFON); Geological Survey Nordrhein Westfalen; Geological Survey Baden-Württemberg; ORFEUS Data Centre; Réseau National de Surveillance Sismique, France; Royal Observatory of Belgium (Belgian and Luxembourg National Networks); Seismological Division KMNI, Netherlands; Seismological Station Bensberg, University of Cologne. Figs. 2-7 were produced with GMT (Wessel and Smith 1998). The Eifel plume project was supported by the Deutsche Forschungsgemeinschaft through grant Ch77/9-4.

References

Ahrens TJ (ed) 1995. A handbook of physical constants I. American Geophys Union, Washington
Albers M, Christensen UR (1996) The excess temperature of plumes rising from the core-mantle boundary. Geophys Res Lett 23:3567-3570
Anderson OL (1999) A thermal balancing act. Science 283:1652-1653

Bina CR, Helffrich G (1994) Phase transition Clapeyron slopes and transition zone seismic discontinuity topography. J Geophys Res 99:15853-15860

Brunet D, Yuen DA (2000) Mantle plumes pinched in the transition zone. Earth Planet Sci Lett 178:13-27

Budweg M, Weber M, Bock G (2001) The upper mantle in the Eifel Plume region. Eos Trans. American Geophys Union 82 Fall Meet Suppl F1115

Davies GF (1988) Ocean bathymetry and mantle convection I. Large-scale flow and hotspots. J Geophys Res 93: 10467-10480

Davies GF (1999) Dynamic Earth. Cambridge University Press

Edholm O, Coffin MF (2000) Large igneous provinces and plate tectonics. In: Richards MA, Gordon RG, van der Hilst RD (eds) The history and dynamics of global plate motions. American Geophysical Union, Washington, Geophysical Monograph 121:309-326

Faul UR, Toomey DR, Waff HS (1994) Intergranular basaltic melt is distributed in thin, elongated inclusions. Geophys Res Lett 21:29-32

Foulger GR, Pritchard MJ, Julian BR, Evans JR, Allen RM, Nolet G, Morgan WJ, Bergsson BH, Erlendsson P, Jakobsdóttir S, Ragnarsson G, Stefansson R, Vogfjörd, K (2000) The seismic anomaly beneath Iceland extends down to the mantle transition zone and no deeper. Geophys J Int 142:F1-F5

Fuchs K, von Gehlen K, Mälzer, H, Murawski, H, Semmel A (eds) (1983) Plateau uplift, the Rhenish shield - a case history. Springer-Verlag, Berlin

Goes S, Spakman W, Bijwaard H (1999) A lower mantle source for Central European volcanism. Science 286:1928-1931

Granet M, Stoll G, Dorel J, Achauer U, Poupinet G, Fuchs K (1995a) Massif Central (France): New constraints on the geodynamical evolution from teleseismic tomography. Geophys J Int 121: 33-48

Granet M, Wilson M, Achauer U (1995b) Imaging a mantle plume beneath the French Massif Central. Earth Planet Sci Lett 136:281-296

Griffiths RW, Campell IH (1990) Stirring and structure in mantle plumes. Earth Planet Sci Lett 99: 66-78

Grunewald S, Kind R, Weber M (2001) The upper mantle under Central Europe - indications for the Eifel plume. Geophys J Int 147:590-601

Jackson I (1998) The Earth's mantle. Cambridge University Press

Jackson I, Fitz Gerald JD, Faul UR, Tan BH (2001) Grain-size sensitive seismic-wave attenuation in polycrystalline olivine. J Geophys Res, submitted

Karato S-i (1993) Importance of anelasticity in the interpretation of seismic tomography. Geophys Res Lett 20:1623-1626

Karato S-i, Wu P (1993) Rheology of the upper mantle. Science 260:771-778

Kampfmann W, Berkhemer H (1985) High temperature experiments on the elastic and anelastic behaviour of magmatic rocks. Phys Earth planet Inter 40:223-247

Kaufmann G, Wu P (2002) Glacial isostatic adjustment in Fennoscandia with a three-dimensional viscosity structure as an inverse problem. Earth Planet Sci Lett 197:1-10

Kennett BLN, Engdahl ER (1991) Traveltimes for global earthquake location and phase identification. Geophys J Int 105:429-465

Keyser M (2001) Dreidimensionale Scherwellen-Geschwindigkeitsstruktur des Lithosphären/Asthenosphären-Systems unter der Eifel. Diploma Thesis, Institute for Geophysics, University of Göttingen

Keyser M, Ritter JRR., Jordan M (2002) 3D shear-wave velocity structure of the Eifel plume, Germany. Earth Planet Sci Lett 203:59-82

Kohlstedt DL, Keppler H, Rubie DC (1996) Solubility of water in the α, β and γ phases of $(Mg,Fe)_2SiO_4$. Contrib Mineral Petrol 123:345-357

Li X, Kind R, Priestley K, Sobolev SV, Tilmann F, Yuan X, Weber M (2000) Mapping the Hawaiian plume conduit with converted seismic waves. Nature 405:938-941

Malamud BD, Turcotte DL (1999) How many plumes are there? Earth Planet Sci Lett 174:113-124

Marquart G, Schmeling H (2000) Interaction of small plumes with the spinel-perovskite phase boundary: implications for chemical mixing. Earth Planet Sci Lett 177:241-254

McKenzie D, Bickle MJ (1988) The volume and composition of melt generated by extension of the lithosphere. J Petrol 29:625-679

Meyer R (2001) Teleseismische P-Wellendämpfung in der Eifel: Analyse und Tomographie. Diploma Thesis, Institute for Geophysics, University of Göttingen

Meyer W, Stets J (1998) Junge Tektonik im Rheinischen Schiefergebirge und ihre Quantifizierung. Z dt geol Ges 149:359-379

Morgan WJ (1971) Convection plumes in the lower mantle. Nature 230:42-43

Nataf H-C (2000) Seismic imaging of mantle plumes. Ann Rev Earth Planet Sci 28:391-417

Ribe NM, Christensen UR (1999) The dynamical origin of Hawaiian volcanism. Earth Planet Sci Lett 171:517-531

Ritter JRR, Christensen UR, Achauer U, Bahr K, Weber M (1998a) Search for a mantle plume under Central Europe. Eos Trans. American Geophys Union 79:420

Ritter JRR, Shapiro SA, Schechinger B (1998b) Scattering parameters in the lithosphere below the Massif Central, France, from teleseismic P-wavefield records. Geophys J Int 134:187-198

Ritter JRR, Achauer U, Christensen UR, the Eifel Plume Team (2000) The teleseismic tomography experiment in the Eifel region, Central Europe: Design and first results. Seism Res Lett 71:437-443

Ritter JRR, Jordan M, Christensen UR, Achauer U (2001) A mantle plume below the Eifel volcanic fields, Germany. Earth Planet Sci Lett 186:7-14

Rothert E, Ritter JRR (2000) Scattering of teleseismic waves under the Eifel region. In: Bonatz M (ed) Comptes-Rendus Journées Luxembourgeoises de Géodynamique, 87:78-81

Sato H, Sacks IS, Murase T, Muncill G, Fukuyama H (1989) Qp-melting temperature relation in peridotite at high pressure and temperature: attenuation mechanism and implications for mechanical properties of the upper mantle. J Geophys. Res 94:10647-10661

Shen Y, Solomon SC, Bjarnason IT, Wolfe CJ (1998) Seismic evidence for a lo-
    wer-mantle origin of the Iceland plume. Nature 395:62-65
Shen Y, Solomon SC, Bjarnason IT, Nolet G, Morgan WJ, Allen RM, Vogfjörd K,
    Jakobsdóttir S, Stefánsson R, Julian BR, Foulger GR (2002) Seismic evidence
    for a tilted mantle plume and north-south mantle flow beneath Iceland. Earth
    Planet Sci Lett 197:261-272
Sleep NH (1990) Hotspots and mantle plumes: Some Phenomenology. J Geophys
    Res 95:6715-6736
Sobolev SV, Zeyen H, Granet M, Achauer U, Bauer C, Werling F, Altherr R,
    Fuchs K (1997) Upper mantle temperatures and lithosphere-asthenosphere
    system beneath the French Massif Central constrained by seismic, gravity,
    petrologic and thermal observations. In: Fuchs K, Altherr R, Müller B,
    Prodehl C (eds) Stress and stress release in the lithosphere – structure and dy-
    namic processes in the rifts of Western Europe, Tectonophysics 275:143-164
Steinberger B, O'Connell RJ (1998) Advection of plumes in mantle flow: implica-
    tions for hotspot motion, mantle viscosity and plume distribution. Geophys J.
    Int 132: 412-434
Tan BH, Jackson I, Fitz Gerald JD (2001) High-temperature viscoelasticity of fi-
    ne-grained polycrystalline olivine. Phys Chem Min 28:641-664
Wessel P, Lyons S (1997) Distribution of large Pacific seamounts from
    Geosat/ERS-1: Implications for the history of intraplate volcanism. J Geophys
    Res 102:22459-22475
Wessel P, Smith WHF (1998) New, improved version of Generic Mapping Tools
    released. Eos Trans. American Geophys Union 79: 579
White RS (1993) Melt production rates in mantel plumes. Phil Trans R Soc Lond.
    A 342:137-153
Wilson JT (1963) A possible origin of the Hawaiian islands. Can J Phys 41:863-
    870
Wolfe CJ, Bjarnason IT, VanDecar JC, Solomon SC (1997) Seismic structure of
    the Iceland mantle plume. Nature 385:245-247
Wörner G (1998) Quaternary Eifel volcanism, its mantle sources and effect on the
    crust of the Rhenish Shield. In: Neugebauer HJ (ed) Young tectonics – mag-
    matism – fluids: a case study of the Rhenish Massif, University of Bonn, SFB
    350 74:11-16
Zhao D (2001) Seismic structure and origin of hotspots and mantle plumes. Earth
    Planet Sci Lett 192:251-265
Zhong S, Watts AB (2002) Constraints on the dynamics of mantle plumes from
    uplift of the Hawaiian Islands. Earth Planet Sci Lett 203:105-116

Structure of the Upper Mantle Beneath Northern Eurasia Derived from Russian Deep-Seismic PNE Profiles

Trond Ryberg[1], Marc Tittgemeyer[2], and Friedemann Wenzel[3]

[1] GeoForschungsZentrum Potsdam, Telegrafenberg, 14473 Potsdam, Germany, E-mail: trond@gfz-potsdam.de
[2] Max-Planck-Institute of Cognitive NeuroScience, Stephanstr. 1a, 04103 Leipzig, Germany
[3] Geophysikaloisches Institut, Universität Karlsruhe, Hertstr. 16, 76187 Karlsruhe, Germany

Abstract

From 1968 until 1990, Russian scientists carried out an intensive program of deep seismic sounding across the territory of the former Soviet Union, using Peaceful Nuclear Explosions (PNEs) as powerful sources for elastic waves. The explosions, both chemical and nuclear, were recorded by up to 400 shot-period (1-2 Hz) three-component, analog recording systems. The average station spacing of about 10 km along the profiles provided a data density not previously available for studies of the upper mantle and the transition zone. Observation distances of more than 3000 km allow the investigation of the velocity structure of the Earth's crust and upper mantle to a depth of 700 km. The analog data have been digitized and used to constrain the fine structure of the upper mantle below Northern Eurasia. They reveal reflections and refractions from upper mantle discontinuities at 410, 520 and 660 km depth. Several properties of the recorded phases have been used to derive a regional P wave velocity model. Synthetic seismograms were calculated and compared with the observations to test these models. Characteristic for all the data in northern Eurasia is the absence of strong pre-critical reflections predicted by the global IASP91 model for the 660 km discontinuity. The appearance of two additional characteristic travel-time branches in the distance range of 2200 km was interpreted as being caused by the proposed and disputed upper mantle discontinuity at 520 km depth. Synthetic seismograms were calculated to constrain its properties. The recordings on the Quartz profile in Northern Eurasia have been used to constrain the nature of the globally observed high-frequency teleseismic P_n phase, which can be observed for shot-receiver distances of more than 3000 km. We suggest that this phase is caused by velocity fluctuations in the upper mantle acting as scatterers. This hypothesis was tested by extensive numerical simulations of the wave propagation using finite difference methods. The present paper is an overview of an extended cooperative study of the upper mantle carried out with PNE seismic data. Although many studies have been carried out in the past, several major topics, for instance the S-wave structure, distribution of attenuation and anisotropy, need further investigations.

1. Introduction: PNE Deep Seismic Sounding in Russia

From 1968 until 1990 an intensive program of seismic investigations using Deep Seismic Sounding (DSS) methods was carried out on the territory of the former Soviet Union. The dense system of DSS profiles covers most of the major tectonic units of northern Eurasia (Fig. 1).

Along these profiles, the elastic waves, generated by chemical underground and Peaceful Nuclear Explosions (PNE) were recorded by three-component portable analog seismic recording systems equipped with short-period seismometers (1-2 Hz). The analog data have been digitized and a data base of deep seismic soundings from PNEs has been created. Waves from the PNEs and conventional chemical explosions recorded along very long DSS profiles (up to 4500 km) penetrate the Earth's upper mantle and mantle transition zone down to 700 km depth and provide velocity and attenuation information about the lithosphere, asthenosphere and the well known velocity discontinuities in the mantle transition zone at 410 and 660 km depth.

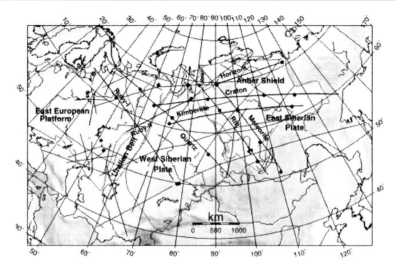

Fig. 1. Map of northern Eurasia. Lines indicate the deep seismic sounding profiles, circles the locations of the peaceful nuclear explosions. Data from profiles with thick lines have been digitized and used for research. Major tectonic provinces are shown.

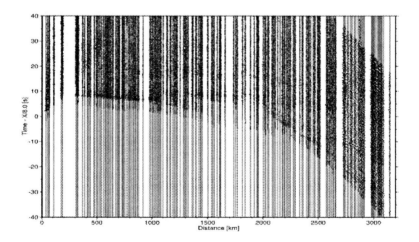

Fig. 2. Example of the seismic data recorded along the profile Quartz from the southernmost shot-point to the northwest. The time-distance record section, which shows the vertical component of ground motion (velocity), is trace normalized and a reduction velocity of 8 km/s has been used. Note that the start of the digitization of the traces is approximately 3-5 seconds before the first arrival. Several seismic phases (triplications) caused by the upper mantle discontinuities can be seen.

The long-range Quartz profile (or Murmansk-Kyzyl profile) is one of the longest (~ 4000 km) DSS profiles in northern Eurasia. It crosses several tectonic units from northwest to southeast including the East European Platform, the Uralian Belt, the West Siberian Platform and the Altai-Sayan Folded Region. Three PNEs and 80 chemical explosions were distributed evenly along the profile. About 400 recording instruments were used for observing the seismic wave field. The maximum observation distance was 3150 km, which corresponds to a penetration depth of about 700 km for refracted waves. As an example, the trace normalized record section of the vertical component of ground velocity from the southernmost shot-point recorded to the northwest is presented in Fig.2. Beyond 300 km distance the first arrivals, which represent the refracted wave through the uppermost mantle, are easily recognized.A common feature of all PNE records is the rapid decay of first arrival amplitudes at distances larger than 1500 km. This is due to a prominent low velocity layer associated with the asthenosphere. At distances greater than 1800 km the wave field is mainly characterized by phases reflected and refracted from the mantle transition zone discontinuities. Earlier overviews of PNE data and their interpretation can be found in Egorkin and Pavlenkova (1981); Egorkin and Chernyshov (1983); Pavlenkova and Egorkin (1983); Egorkin et al. (1987); Egorkin and Mikhaltsev (1990); Mechie et al. (1993) and Ryberg et al. (1996).

2. The Upper Mantle Discontinuities at 660 and 520 km Depth

Seismological models of upper mantle structure down to the depths of the 410 km and 660 km discontinuities provide important constraints on the composition and dynamics of the mantle. There is general agreement that the 410 km discontinuity is associated with a phase transition from olivine to β spinel (Ito and Takahashi, 1989), although the amount of olivine constituting the upper mantle remains unclear and may vary from 40% (Duffy and Anderson, 1989; Duffy et al. 1995) to 60% Irifune, 1987; Bina, 1993). The nature of the 660 km discontinuity is even more controversial. The velocity increase may be due to a transition from γ spinel to perovskite and magnesiowüstite and/or a transition from garnet to perovskite. Both components probably to contribute to the transition. These two transitions may actually occur at slightly different depths or over different depth intervals, so that the 660 km transition may be more complex than just a single gradient or a single discontinuity (Ito and Takahashi, 1989; Duffy and Anderson, 1989; Irifune, 1987; Kuskov and Panferov, 1991; Ita and Stixrude,

1992). For years a debate was focused on whether the transition zone separates material of different chemistry. Different chemistries of the upper and lower mantle would indicate that mantle convection occurs as two-layer convection, whereas a simple phase transition would argue for whole-mantle convection. Modern tomographic studies of subduction zones (Creager and Jordan; 1984; Creager and Jordan, 1986; van der Hilst et al., 1991; Fukao et al., 1992; Widiyantoro and van der Hilst, 1996; van der Hilst et al., 1997; Bijwaard et al., 1098; Ritsema et al., 1999) show that at least for some subduction zones the convection process is not limited to the upper mantle.

To investigate the properties of the mantle transition zone below northern Eurasia, we analyzed seismic data from 18 shots recorded on seven profiles for the 410 and 660 km discontinuities. An example of a vertical component PNE record section is given in Figure 2. Low-pass filtering with a corner frequency of 1.25 Hz was applied to the data to reduce the high-frequency noise. Only the distance range where we expect the triplications from the 410 and 660 km discontinuities was analyzed. Figure 3a shows the stacking result focussing on the 660 km discontinuity. We stacked the envelopes of all available vertical component PNE data (thick lines in Figure 1). Common to almost all PNE record sections, especially for observations along the westernmost profiles, is the amplitude decay of the first arrivals from 1700 to 2100 km associated with a low velocity zone in the upper mantle (absence of energy associated with the refracted P phase).

While the seismic phases associated with the triplication from the 410 km discontinuity are in agreement with the predictions by global earth models the corresponding phases from the 660 km discontinuity differ systematically from the predictions. Figure 3b presents the theoretical travel-time curves for the refracted and postcritically reflected phases associated with the mantle transition zone based on the standard Earth model IASP91 (Kennett and Engdahl, 1991). To compare our stacking results with the wave field predicted by IASP91, synthetic seismograms were calculated using the reflectivity method (Fuchs and Müller, 1971). To allow for a simple comparison of the observed data with the synthetics, we applied the same stacking procedure to the synthetically calculated data (Figure 3c). The IASP91 model predicts large amplitudes for these phases (Figure 3c), with the critical point for reflections from the 660 km discontinuity located at 1967 km which is not observed in the data. The second discrepancy between IASP91 and our data can be observed at 3000 km distance. IASP91 predicts a strong postcritical reflection beyond this offset. The data suggest its termination at 3000 km. Two reasons may be responsible for the weak

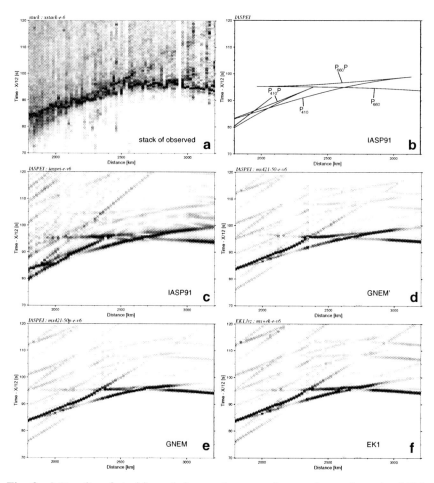

Fig. 3. a) Results of stacking of observations to enhance phases from the 660 km discontinuity (modified from Fig. 5, (Ryberg et al., 1998). b) gives an overview of the expected travel times of the upper mantle phases according to the IASP91 model. The calculations for IASP91 are shown in c. Note the strong critical reflection from the 660 km discontinuity at around 2000 km distance for IASP91. e) shows the stacking result for our preferred Generalized Northern Eurasia Model (GNEM), (Ryberg et al., 1998). It has a reduced P wave velocity step across the 660 km discontinuity and agrees with the observations of the PNEs. The stacking result for an alternative model with a 50 km wide transitional 660 km discontinuity (d) is also in agreement with the observations. f) was calculated for the model EK1 proposed by Estabrook and Kind (1996) and shows also the absence of strong critical 660 km reflections.

precritical energy of the 660 km phase. First, it could be caused by a reflection at a discontinuity with a reduced P wave velocity step across it, which would shift the critical point to a larger distance. Figure 3e shows the synthetics of a model with a reduced (50% of IASP91) velocity step across the boundary (Generalized Northern Eurasia Model (GNEM)), and figure 3f shows those of the model EK1 (modified to have an upper mantle low-velocity zone added) by Estabrook and Kind, (1996). Alternatively, the lack of critical energy around 2000 km could be explained by a deeper located 660 km discontinuity (labeled GNEM' in Figure 3d).

Recent studies of the upper mantle concentrate on fine structural features such as the sharpness of discontinuities, their lateral variations, slab-mantle interaction and structures within the transition zone as, for instance, a possible 520 km discontinuity. There are several studies that suggest a velocity discontinuity in the depth range from 500 to 550 km (Wiggins and Helmberger, 1973; Simpson et al., 1974; Helmberger and Engen, 1974; Hales et al. 1980). During the last few years, new data from global and regional arrays became available and revived the discussion. Shearer (1990, 1991) and Revenaugh and Jordan (1991) found evidence for the discontinuity in long-period data, although Bock (1994) questioned the significance of the data. Jones et al. (1992) used the short-period Southern California Seismic Network to observe Latin American earthquakes. Cummins et al. (1992) observed many events from the Indonesian Arc with a sparse mobile Australian array and combined them into a record section. Whereas Jones et al. (1992) and Cummins et al. (1992) looked at wide-angle refracted waves, Benz and Vidale (1993) scrutinized P'P' phases where waves impinge almost vertically on the upper mantle discontinuities. Figure 4 (top) shows the record section of Quartz-3 (zoom of Figure 2). It contains the clearest indication for an additional upper mantle discontinuity. In figure 4 (top) several branches of the travel time curve are present. From 2200 to 2275 km distance the first arrival has an apparent velocity of 10.25 km/s and represents the wave refracted below the 410 km discontinuity. The first arrivals for distances greater than 2700 km have an apparent velocity of 12.2 km/s and are caused by waves refracted below the 660 km discontinuity. In the distance range from 2300 to 2700 km the apparent velocity of the first arrivals is about 11 km/s. The first arrivals have a relatively low amplitude and are followed by a secondary phase with significantly higher amplitudes. The high-amplitude secondary phase has a lower apparent velocity of 10.4 km/s. The distance at which a low-amplitude phase appears as the first arrival is approximately 2400 km.

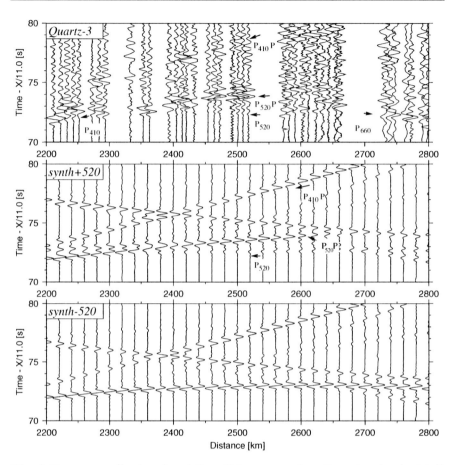

Fig. 4. Trace normalized, reduced time-distance record sections showing in detail the seismic phases associated with the 520 km discontinuity in the observed data for the Quartz profile (top), the best fitting model (middle), and a model without the 520 km discontinuity (bottom) (Fig. 6, (Ryberg et al. 1997). The refracted phases P_{410}, P_{520} and P_{660}, and the reflected phases $P_{520}P$ and $P_{410}P$ are marked by arrows.

To study the change in properties at the 520 km discontinuity, we calculated synthetic seismograms for different one-dimensional models, using the reflectivity method. The model with a velocity jump across the 520 km boundary of 0.25 km/s and a gradient below this boundary of 0.001 1/s produces a wave field with amplitudes and apparent velocities similar to the observed ones (Ryberg et al., 1997). The comparison with the recordings from the Quartz profile shows a similar behavior of the phases related to the 520 km discontinuity, and therefore, provides evidence for the existence of this boundary.

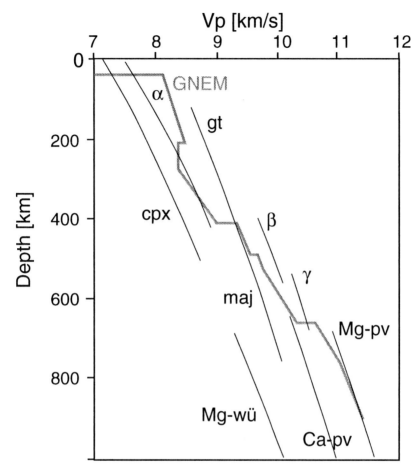

Fig. 5. Comparison of the model GNEM derived from PNE recordings with the expected P wave velocities for different mantle rocks. The velocity-depth functions of α-, β- and γ-phases of olivine, cpx clinopyroxene, Magnesio- and Calcium-perovskite (Mg-pv and Ca-pv), Magnesiowütite (Mg-wu), Majorite and Garnet (gt) are shown (modified from Bass and Anderson (1984). The upper mantle model GNEM is characterized by a low-velocity zone starting at 210 km depth and a reduced velocity step (with respect to IASP91) across the 660 km discontinuity.

There are different hypotheses as to the nature of the 520 km discontinuity. While Katsura and Ito (1989) suggested a model based on a phase transition from β olivine to γ spinel at this depth, Revenaugh and Jordan (1991) proposed models where the dissolution of garnet to Ca-perovskite causes a positive velocity contrast. Other models (Gasparik, 1993) suggest that the 520 km discontinuity is related to the base of a garnet layer. The velocity increase of 0.25 km/s is only slightly higher than that expected for the β olivine to γ spinel phase transition (Duffy and Anderson, 1989). Thus a very olivine rich upper mantle could explain the velocity jump. However, the amount of olivine in the upper mantle transition zone is still a matter of debate. Duffy and Anderson (1989) rule out an olivine content larger than 50%. Their preferred models contain 40% olivine. Bina (1993), on the other hand, claims that values of 40% to 70% are conceivable.

Summarizing our studies, figure 5 shows a generalized P-wave velocity model (GNEM) for Northern Eurasia with an uppermost mantle low velocity zone, a 660 km discontinuity with a decreased velocity contrast and the additional 520 km discontinuity.

3. High-Frequency Scattering in the Uppermost Mantle

Mantle P_n and S_n phases with unusually high frequencies and long codas from earthquakes and other sources have been observed in many different areas of the world (e.g., Linehan, 1940; Molnar and Oliver, 1969; Walker, 1977) as early as 1935. These phases are called oceanic P_n, P_O and long range or teleseismic P_n, and are characterized by group velocities of ~ 8.0 km/s. It is remarkable that the high-frequency signals (typically >15 Hz, occasionally as high as 35 Hz) are reported for both continental and oceanic paths. They travel efficiently across different tectonic provinces such as continental shields and deep-ocean basins (Molnar and Oliver, 1969), and appear to be interrupted only by major plate tectonic boundaries, such as mid-ocean ridges, island arc structures and subduction zones.

Although many theories have been advanced to explain these waves, no commonly accepted waveguide mechanism for their efficient propagation exists. Several explanations for high-frequency teleseismic P_n and S_n have been proposed: whispering gallery waves multiply reflected at the crust-mantle boundary; transmission in a high-velocity layer beneath the Moho and above a low-velocity channel; upper mantle "lid" modes; tunneling of low-frequency waves through thin high-velocity layers; models with re-verberations in the lithosphere, crust and water column; transmission through a low-velocity zone below the Moho. For a detailed overview and alternative models see Sereno and Orcutt (1985, 1987); Brandsdottir and Menke (1988); Mallick and Frazer (1990); Mantovani et al. (1977); Gettrust and Frazer (1981); Menke and Richards (1980); Sutton and Walker (1972); Stephens and Isacks (1977). Models with random velocity heterogeneities in the crust or upper mantle have been suggested and studied by Menke and Chen (1984); Richards and Menke (1983); Mallick and Frazer

(1990) and Tittgemeyer et al. (1996). Simulations of wave propagation by analog model studies Menke and Richards (1983) show that scattering effects in the upper mantle could explain the coda phenomenon of the teleseismic P_n phase.

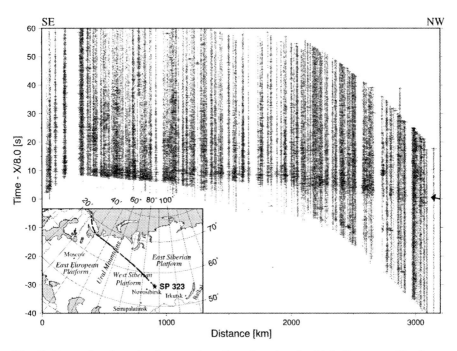

Fig. 6. Example of a seismic record section showing the high-frequency P_n phase. The vertical component record section from the long-range seismic profile Quartz was high-pass filtered (corner frequency 5Hz). A strong band of energy traveling with a group velocity of $V = 8.1$ km/s is the high-frequency teleseismic P_n phase.

Figure 6 shows the remarkable pattern of the high-frequency wave field recorded along the PNE profile Quartz. In contrast to the low-frequency part of the wave field, there are only weak phases from the mantle. The entire section is dominated by a band of seismic energy arriving as a strong secondary phase with a group velocity of about 8.1 km/s (teleseismic P_n phase). This P_n phase is apparent in the record section from a few hundred kilometers to the maximum recording distance of 3145 km. At a distance of about 750 km, it separates from the faster first arrivals. In this filtered (high-pass) section, it becomes the strongest secondary phase, and its amplitude exceeds that of the first arrival beyond 1400 km. This high-frequency phase has no recognizable sharp onset and is spatially incoherent. A remarkable property of the phase is the long, pronounced coda. The same characteristic high-frequency feature is observed in the record sections of both the radial and transverse components of the P wave window.

It is commonly accepted that scattering at small-scale heterogeneities in the crust and upper mantle associated with two- and three-dimensional velocity and density fluctuations is responsible for most of the seismic coda. To explain both the scattering and the teleseismic propagation, we propose a velocity model containing small-scale velocity fluctuations in the upper mantle below the Moho (Figure 7).

The aim of our investigation was to verify whether two-dimensional velocity fluctuations in the upper mantle could explain the observation of high-frequency P_n phases. We performed a grid search to scan the possible range of layer thickness, L, heterogeneity correlation lengths, a_x and a_z, standard deviation of heterogeneity (magnitude of fluctuations with respect to background model), σ, and other properties (V_p-V_s correlation, background model, random distribution functions, etc.). We used finite difference techniques (Kelly et al. 1976) to calculate the elastic wave field in complex models. We started our studies with the analysis of the wave propagation in the background model IASP91. Fig. 8b shows the unfiltered, vertical component record section for the background model. The record section is mainly characterized by the Moho reflection (P_MP), and the phase diving into the upper mantle (P). The secondary wave field consists of their respective multiple reflected and converted waves generated at the free surface (P') and the Moho. The mantle refraction remains weak between the critical angle of P_MP and 750 km. It splits into two phases at a distance of around 1000 km, withthe fast phase being the refracted wave diving deeper into the upper mantle. The second phase has a lower apparent velocity (~ 8 km/s) and represents the branches of the whispering-gallery phase (WG) at the crust/mantle boundary. This phase was interpreted by Morozov et al. (1998) to cause the high-frequency P_n phase. The whispering-gallery phase, although having arrival times and an apparent velocity similar to the observations, has a different frequency character and is not pursued by a long coda. No feature like the teleseismic P_n with its pronounced coda is visible. The background model and its synthetic seismograms serve as a reference for further calculations, which include mantle velocity fluctuations.

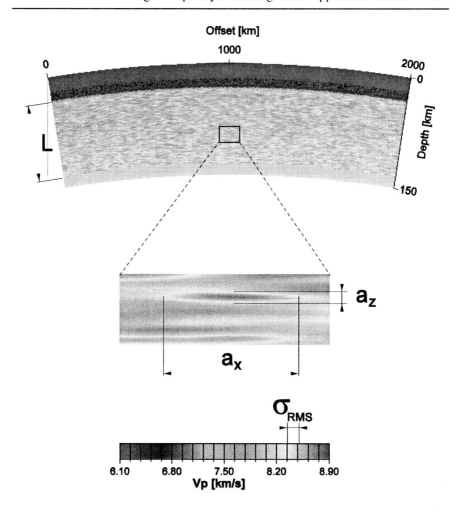

Fig. 7. Long distance wave propagation of P_n is explained by a velocity model characterized by statistical fluctuations of the velocity in the uppermost mantle to a depth of $\sim$ 130km (from Tittgemeyer et al. 2002). Random mantle heterogeneities have a Gaussian distribution with a horizontal (a_x) and vertical (a_z) correlation length of 20 km and 0.5 km, respectively. The standard deviation (σ) of the velocity fluctuations is 2 %. L is the thickness of the layer containing the velocity fluctuations.

By perturbing the smooth upper mantle gradient with weak velocity fluctuations the wave field changes significantly (see Fig. 8d). As σ increases to more than 1 %, the main energy is travels as a secondary phase: the high-frequency teleseismic P_n. A teleseismic P_n separates from the refracted mantle phase P at distances $\geq$ 1000 km. While the latter phase is

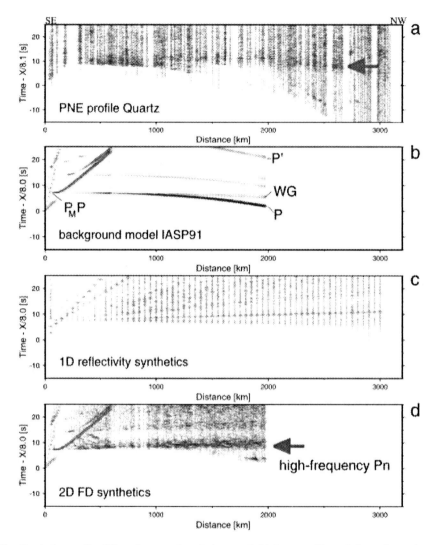

Fig. 8. a) shows the filtered example of observed, high-pass filtered data along the Russian long-range profile Quartz. b) Synthetic seismograms computed for the reference model IASP91 (without scatterers) (Kennett and Engdahl, 1991). This model is characterized by a simple and smooth velocity structure. c) Introducing into the IASP91 model upper mantle velocity fluctuations (1D model with lamellae-like structures) we calculated synthetic seismograms with the reflectivity method. The synthetics start to show a high-frequency phase. d) shows the wave field calculated by FD methods for the two-dimensional scattering model of Figure 7. Note the similarity of the marked high-frequency phase to a).

low frequency, the teleseismic P_n is dominated by a higher frequency content. This behaviour is similar to the observed data (Fig. 8a) and is shown to be a characteristic feature of teleseismic P_n observations (Ryberg et al., 1995; Tittgemeyer et al., 1997).

The comparison of wave propagation in different models with variable parameters of random velocity fluctuation led us to conclude that these models can explain the propagation of a high-frequency teleseismic P_n. It could shown that only models with specific properties of the velocity structure are capable of propagating these phases efficiently, (Ryberg et al., 2000; Ryberg and Wenzel, 1999; Tittgemeyer et al., 1999; Tittgemeyer et al, 2000). A small, but significant amount of velocity fluctuations of 1.5 to 2 %, a predominantly elongated structure with a_x and a_z around 20 and 0.5 km, respectively, and a sufficiently thick heterogeneous zone ($L \geq 100$ km) are necessary to produce synthetics similar to the observations.

Mallick and Frazer (1990) discussed the variability in upper mantle composition required to generate a few percent velocity variation. Conceivable variations in the olivine, ortho- and clinopyroxene and garnet content of both typical peridotitic and eclogitic mantle materials can easily explain variations of the P and S velocities of 2 %. However, a regional study in the French Massif Central by Sobolev et al. (1997), using a xenolith database, shows only small (0.5 %) variations for P and S wave velocities. Hence, as an alternative to the suggested upper mantle model with randomly varying compositions we propose a model based on spatially varying intrinsic anisotropy. In this model heterogeneity is represented by spatial domains of intrinsic anisotropy (e.g. a variability of the mean anisotropic symmetry orientation). Within these domains the fast symmetry axis (a [100]) is preferentially horizontally oriented. This orientation differs from domain to domain by small angles in a random fashion. The respective b and c axes are assumed to be randomly oriented. This model resembles an idea published by Fuchs (1977).

Acknowledgements

We thank J. Ansorge, J. Ritter and J. Mechie for helpful comments, corrections and suggestions that improved the manuscript. The digital data for the PNE profiles were kindly made available by the GEON/Russian Geological Committee within the EUROPROBE cooperation agreement. The Ministry of Research and Technology of Germany (BMBF) supported the digitization of the PNE data at GEON. The research has been partly funded by the Deutsche Forschungsgemeinschaft (grant WE 1394/2-1), the Geo-ForschungsZentrum Potsdam and the BMBF (grant RG 9216).

References

J.D. Bass and D.L. Anderson. Composition of the upper mantle: Geophysical tests of two petrological models.*Geophys. Res. Lett.*, 11:237-240, 1984.

H.M. Benz and J.E. Vidale. The sharpness of upper mantle discontinuities determined from high-frequency reflections. *Nature*, 365:147-150, 1993.

H. Bijwaard, W. Spakman, and E.R. Engdahl. Closing the gap between regional and global travel time tomography. *J. Geophys. Res.*, 103:30,055-30,078, 1998.

C.R. Bina. Mutually consistent estimates of upper mantle composition from seismic velocity contrasts at 400 km depth. *Pure Appl. Geophys.*, 141:101-109, 1993.

G. Bock. Synthetic seismogram images of upper mantle structure: No evidence for a 520-km discontinuity. *J. Geophys. Res.*, 99:15.843-15.851, 1994.

B. Brandsdóttir and W.H. Menke. Measurements of coda buildup and decay rates of Western Pacific P, P_o, and S_o phases and their relevance to lithospheric scattering. *J. Geophys. Res.*, 93:10.541-10.559, 1988.

K.C. Creager and T.H. Jordan. Slab penetration into the lower mantle. *J. Geophys. Res.*, 89:3.031-3.049, 1984.

K.C. Creager and T.H. Jordan. Slab penetration into the lower mantle beneath the Mariana and other island arcs of the northwest Pacific. *J. Geophys. Res.*, 91:3.573-3.589, 1986.

P.R. Cummins, B.L.M. Kennett, J.R. Bowman, and M.G. Bostock. The 520 km discontinuity?
Bull. Seismol. Soc. Am., 82:323-336, 1992.

T.S. Duffy and D.L. Anderson. Seismic velocities in mantle minerals and the mineralogy of the upper mantle. *J. Geophys. Res.*, 94:1.895-1.912, 1989.

T.S. Duffy, C.S. Zah, R.T. Downs, H.K. Mao, and R.J. Hemley. Elasticity of forsterite to 16 GPa and the composition of the upper mantle. *Nature*, 378:170-173, 1995.

A.V. Egorkin and N.M. Chernyshov. Peculiarities of mantle waves from long-range profiles. *J. Geophys.*, 54:30-34, 1983.

A.V. Egorkin and A.V. Mikhaltsev. The result of seismic investigations along geotraverses. In K. Fuchs, Y.A. Kozlovsky, A.I. Krivtsov, and M.D. Zoback, editors, Super-Deep Continental Drilling and Deep Geophysical Sounding, pages 111-119. Springer Verlag Berlin Heidelberg New York, 1990.

A.V. Egorkin and N.I. Pavlenkova. Studies of mantle structure of U.S.S.R. territory on long-range seismic profiles. *Phys. Earth Planet. Inter.*, 25:12-26, 1981.

A.V. Egorkin, S.K. Zuganov, N.I. Pavlenkova, and N.M. Chernyshov. Results of lithospheric studies from long-range profiles in Siberia. *Tectonophysics*, 140:29-47, 1987.

C.H. Estabrook and R. Kind. The nature of the 660 kilometer upper-mantle seismic discontinuity from precursors to the PP phase. *Science*, 274:1.179-1.182, 1996.

K. Fuchs. Seismic anisotropy in the subcrustal lithosphere as evidence for dynamical processes in the upper mantle. *Geophys. J. R. Astr. Soc.*, 49:167-179, 1977.

K. Fuchs and G. Müller. Computation of synthetic seismograms with the reflectivity method and comparison with observations. *Geophys. J. R. Astr. Soc.*, 23:417-433, 1971.

Y. Fukao, M. Obayashi, H. Inoue, and M. Nenbai. Subducting slabs stagnant in the mantle transition zone. *J. Geophys. Res.*, 97:4.809-4.822, 1992.

T. Gasparik. The role of volatiles in the transition zone. *J. Geophys. Res.*, 98:4.287-4.299, 1993.

J.F. Gettrust and L.N. Frazer. A computer model study of the propagation of long-range P_n phase. *Geophys. Res. Lett.*, 8:749-752, 1981.

A.L. Hales, K.J. Muirhead, and J.R. Rynn. A compressional velocity distribution for the upper mantle. *Tectonophysics*, 63:309-348, 1980.

D.V. Helmberger and G.~R. Engen. Upper mantle shear structure. *J. Geophys. Res.*, 79:4.017-4.028, 1974.

T. Irifune. An experimental investigation of the pyroxene-garnet transformation in a pyrolite composition and its bearing on the constitution of the mantle. *Phys. Earth Planet. Inter.*, 45:324-336, 1987.

J. Ita and L. Stixrude. Petrology, elasticity, and composition of the mantle transition zone. *J. Geophys. Res.*, 97:6.849-6.866, 1992.

E. Ito and E. Takahashi. Postspinel transformations in the system Mg_2SiO_4-Fe_2SiO_4 and some geophysical implications. *J. Geophys. Res.*, 94:10.637-10.646, 1989.

L.E. Jones, J. Mori, and D.V. Helmberger. Short-period constraints on the proposed transition zone discontinuity. *J. Geophys. Res.*, 97:8.765-8.774, 1992.

T. Katsura and E. Ito. The system Mg_2SiO_4-Fe_2SiO_4 at high pressures and temperatures: Precise determination of stabilities of olivine, modified spinel, and spinel. *J. Geophys. Res.*, 94:15.663-15.670, 1989.

K.R. Kelly, R.W. Ward, S. Treitel, and R. M. Alford. Synthetic seismograms: a finite-difference approach. *Geophysics*, 41:2-27, 1976.

B.L.N. Kennett and E.R. Engdahl. Traveltimes for global earthquake location and phase identification. *Geophys. J. Int.*, 105:429-465, 1991.

O.L. Kuskov and A.B. Panferov. Constitution of the mantle, 3, Density, elastic properties and the mineralogy of the 400 km discontinuity. *Phys. Earth Planet. Inter.*, 69:85-100, 1991.

D. Linehan. Earthquakes in the West Indian region. *EOS Trans. Am. Geophys. Union*, 21:229-232, 1940.

S. Mallick and L.N. Frazer. P_o/S_o synthetics for a variety of oceanic models and their implications for the structure of the oceanic lithosphere. *Geophys. J. Int.*, 100:235-253, 1990.

E. Mantovani, F. Schwab, H. Liao, and L. Knopoff. Teleseismic S_n: a guided wave in the mantle. *Geophys. J. R. Astr. Soc.*, 51:709-726, 1977.

J. Mechie, A.V. Egorkin, K. Fuchs, T. Ryberg, L. Solodilov, and F. Wenzel. P-wave mantle velocity structure beneath northern Eurasia from long-range recordings along the profile Quartz. *Phys. Earth Planet. Inter.*, 79:269-286, 1993.

W. H. Menke and R. Chen. Numerical studies of the coda falloff rate of multiply-scattered waves in randomly layered media. *Bull. Seismol. Soc. Am.*, 74:1.605-1.621, 1984.

W.H. Menke and P.G. Richards. Crust-mantle whispering gallery phases: A deterministic model of teleseismic P_n wave propagation. *J. Geophys. Res.*, 85:5.416-5.422, 1980.

W.H. Menke and P.G. Richards. The horizontal propagation of P waves through scattering media: Analog model studies relevant to long range P_n propagation. *Bull. Seismol. Soc. Am.*, 73:125-142, 1983.

P. Molnar and J. Oliver. Lateral variations of attenuation in the upper mantle and discontinuities in the lithosphere. *J. Geophys. Res.*, 74:2.648-2.682, 1969.

I.B. Morozov, E.A. Morozova, S.B. Smithson, and L.N. Solodilov. On the nature of the teleseismic P_n phase observed on the ultralong-range profile Quartz, Russia. *Bull. Seismol. Soc. Am.*, 88(No. 1):62-73, 1998.

N.I. Pavlenkova and A.V. Egorkin. Upper mantle heterogeneity in northern part of Eurasia. *Phys. Earth Planet. Inter.*, 33:180-193, 1983.

J. Revenaugh and T.H. Jordan. Mantle layering from ScS reverberations, 2, The transition zone. *J. Geophys. Res.*, 96:19.763-19.780, 1991.

P.G. Richards and W.H. Menke. The apparent attenuation of a scattering medium. *Bull. Seismol. Soc. Am.*, 73:1.005-1.021, 1983.

J. Ritsema, H. Jan van Heijst, and J.H. Woodhouse. Complex shear wave velocity structure imaged beneath Africa and Iceland. *Science*, 286:1925-1931, 1999.

T. Ryberg, K. Fuchs, A.V. Egorkin, and L. Solodilov. Observation of high-frequency teleseismic P_n waves on the long-range Quartz profile across Northern Eurasia. *J. Geophys. Res.*, 100:18.151-18.163, 1995.

T. Ryberg, M. Tittgemeyer, and F. Wenzel. Finite difference modelling of P-wave scattering in the upper mantle. *Geophys. J. Int.*, 141:787-801, 2000.

T. Ryberg and F. Wenzel. High-frequency wave propagation in the uppermost mantle. *J. Geophys. Res.*, 104:10.655-10.666, 1999.

T. Ryberg, F. Wenzel, A.V. Egorkin, and L. Solodilov. Short-period observation of the 520 km discontinuity in northern Eurasia. *J. Geophys. Res.*, 102:5.413--5.422, 1997.

T. Ryberg, F. Wenzel, A.V. Egorkin, and L. Solodilov. Properties of the mantle transition zone in northern Eurasia. J. Geophys. Res., 103:811-822, 1998.

T. Ryberg, F. Wenzel, J. Mechie, A.V. Egorkin, K. Fuchs, and L. Solodilov. 2D-velocity structure beneath northern Eurasia derived from the super long-range seismic profile Quartz. *Bull. Seismol. Soc. Am.*, 86:857-867, 1996.

T.J. Sereno and J.A. Orcutt. Synthesis of realistic oceanic P_n wave trains. *J. Geophys. Res.*, 90:12.755-12.776, 1985.

T.J. Sereno and J.A. Orcutt. Synthetic P_n and S_n phases and the frequency dependence of Q of oceanic lithosphere. *J. Geophys. Res.*, 92:3.541-3.566, 1987.

P.M. Shearer. Seismic imaging of upper-mantle structure with new evidence for a 520 km discontinuity. *Nature*, 344:121-126, 1990.

P.M. Shearer. Constraints on upper-mantle discontinuities from observations of long-period reflected and converted phases. *J. Geophys. Res.*, 96:18.147-18.182, 1991.

D.W. Simpson, R.F. Mereu, and D.W. King. An array study of P-wave velocities in the upper mantle transition zone beneath northeastern Australia. *J. Geophys. Res.*, 64:1.757-1.788, 1974.

S.V. Sobolev, H. Zeyen, M. Granet, U. Achauer, C. Bauer, F. Werling, R. Altherr, and K. Fuchs. Upper mantle temperatures and lithosphere-asthenosphere system beneath the French Massif Central constrained by seismic, gravity, petrologic and thermal observations. *Tectonophysics*, 275:143-164, 1997.

C. Stephens and B.L. Isacks. Toward an understanding of S_n; Normal modes of Love waves in an oceanic structure. *Bull. Seismol. Soc. Am.*, 67:69-78, 1977.

G.H. Sutton and D.A. Walker. Oceanic mantle phases recorded on seismographs in the northwestern Pacific at distances between 7° and 40°. Bull. Seismol. Soc. Am., 62:631-655, 1972.

M. Tittgemeyer, T. Ryberg, K. Fuchs, and F. Wenzel. Observation of teleseismic P_n/S_n on super long-range seismic profiles in northern Eurasia and their implications for the structure of the lithosphere. In K. Fuchs, editor, *Upper Mantle Heterogeneities from Active and Passive Seismology*, NATO ASI Series, pages 63-73. Kluwer Academic Publishers, Dordecht /Boston /London, 1997.

M. Tittgemeyer, T. Ryberg, F. Wenzel, and K. Fuchs. Heterogeneities of the Earth's uppermost mantle. In J.A. Goff and K. Holliger, editors, *Heterogeneity in the Crust and Upper Mantle: Nature, Scaling and Seismic Properties*, pages 281-297. Kluwer Academic Publishers, Dordecht /Boston /London, 2002.

M. Tittgemeyer, F. Wenzel, and K. Fuchs. On the nature of P_n. *J. Geophys. Res.*, 107:16,173-16,180, 2000.

M. Tittgemeyer, F. Wenzel, K. Fuchs, and T. Ryberg. Wave propagation in a multiple-scattering upper mantle - observations and modelling. *Geophys. J. Int.*,127:492-502, 1996.

M. Tittgemeyer, F. Wenzel, T. Ryberg, and K. Fuchs. Scales of heterogeneities in the continental crust and upper mantle. *Pure Appl. Geophys.*, 156:29-52, 1999.

R.D. van der Hilst, E.R. Engdahl, W. Spakman, and G. Nolet. Tomographic imaging of subducted lithosphere below northwest Pacific island arcs. *Nature*, 353:37-43, 1991.

R.D. van der Hilst, S. Widiyantoro, and E.R. Engdahl. Evidence for deep mantle circulation from global tomography. *Nature*, 386:578-584, 1997.

D.A. Walker. High-frequency P_n and S_n phases recorded in the western Pacific. *J. Geophys. Res.*, 82:3.350-3.360, 1977.

S. Widiyantoro and R. van der Hilst. Structure and evolution of lithospheric slab beneath the Sunda Arc, Indonesia. *Science*, 271:1.566-1.570, 1996.

R.A. Wiggins and D.V. Helmberger. Upper mantle structure of the western United States. *J. Geophys. Res.*, 78:1.870-1.880, 1973.

True Amplitudes: A Challenge in Reflection Seismology

Christoph Jäger[1], Thomas Hertweck[1], and Alexander Goertz[2]

[1] Geophysical Institute, University of Karlsruhe, Hertzstr. 16, D-76187 Karlsruhe, Germany
[2] Paulsson Geophysical Services, Inc. (P/GSI), 1215 West Lambert Road, Brea, CA 92821-2819, USA

Abstract

Kirchhoff migration is a well-known process in the world of seismic exploration to transform seismic reflection data into an interpretable image of the subsurface. In former times, only kinematic traveltime information were utilized to image interfaces within the earth. That means, the dynamic information related to the recorded amplitudes remained unused. During the last decades, migration concepts have strongly improved and make nowadays additional use of these dynamic information to obtain knowledge about petrophysical rock properties. By applying suitable weight functions in the migration process the geometrical spreading effects of propagating waves can be compensated. Such an approach is called "true-amplitude migration". In this way, the output amplitudes are related to the reflection coefficient. As a consequence, detailed amplitude versus offset (AVO) or angle (AVA) analysis may be performed and, thus, the search for, e.g., reservoirs is improved.

1. Introduction

In the early days of hydrocarbon exploration by seismic methods, a subsurface "image" was obtained by simply turning the field records upside down so that the time axis of the seismograms pointed downwards. By scaling with the propagation velocity that could be measured from the moveout of the arrival time with distance from the source (offset), the extrapolated arrival time of reflections directly at the source (i.e., at zero offset) was a measure of the depth to the reflector. If carried out for a number of observation points, subsurface reflectors were mapped by their (velocityscaled) zero-offset reflection time along a profile. However, it was soon discovered that this technique leads to inaccurate results as soon as the subsurface differs from being horizontally layered.

It was Hagedoorn (1954) who first discovered a technique to move reflection events recorded in the field and displayed in the above mentioned manner to their correct spatial location: if a half-circle was drawn around the origin of each seismogram (zero-offset configuration) with the radius equal to the one-way traveltime of a reflection event recorded on this trace times the estimated velocity, one observes that all these half-circles have a common envelope that defines the correct location of the 2 Jäger et al. reflector (see Figure 1). By this means, reflection events could be moved (i.e., migrated) to their correct location—the concept of seismic migration was born. Since this technique did not require any additional expenditure and could be carried out directly in the field even without a compass at hand, it has also become known as shoestring migration (see, e.g., Bleistein, 1999).

Physically, each half-circle corresponds to an isochron, i.e., the curve (2-D) or surface (3-D) of equal reflection time in the image space. Hagedoorn called them "curves/surfaces of maximum concavity". Of course, these curves are only halfcircles for the case of zero-offset configuration and a homogeneous model, i.e., a constant propagation velocity of the seismic waves. If the medium velocity varies, the shoestring has to be replaced by rays which are traced through the subsurface until the measured one-way reflection time is reached. With the upcoming of computers, the method was implemented in a slightly different way: to avoid the picking of events in the time domain, all points on a trace were treated as possible reflection events and the amplitudes found at these positions were spread along the corresponding isochrons. This approach relies on the constructive and destructive interference of all the isochrons in the image space and is nothing else than a systematic realization of Huygens' principle. Some image points get a large amplitude value as migration result and, thus, describe the location of reflectors within the earth. The resulting amplitude for other image points is negligible due to the destructive interference of all contributing isochrons.

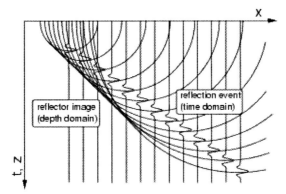

Fig. 1. Hagedoorn's principle of seismic migration, also called shoestring migration. The correct location of a reflector in depth is defined as the envelope of all isochrons belonging to the same reflection event in the time domain. As can be seen, the location of the reflector image as well as its dip and length change after migration.

This migration method can also be realized in a different way: instead of distributing reflection event amplitudes along isochrons, we may compute the diffraction traveltimes for every point in the image space. Then, we sum up the amplitudes encountered in the seismic time section along each diffraction curve (2-D) or surface (3-D) and assign the value to the respective image point. Both migration schemes are True amplitudes: a challenge in reflection seismology 3 conceptually equivalent, only the order of the summation steps is interchanged. The close relationship between the isochron and the diffraction traveltime curve/surface, which is also called Huygens curve/surface, has become known as the concept of duality (Tygel et al., 1995). In the world of seismic exploration, a summation of amplitudes is always called a stack. Hence, the latter described method of using the diffraction traveltime curves/surfaces is called diffraction-stack migration, as opposed to the isochron algorithm which is known as smearing method.

In mathematical terms, this summation can be described by an integral (Schneider, 1978) which is called the diffraction-stack integral. Because of its great resemblance to the classic Kirchhoff integral representation of the solution of the wave equation, the term Kirchhoff migration or Kirchhoff-type migration has become a commonly accepted but unfortunately slightly misleading terminology for the method. Since the Kirchhoff integral by itself cannot be used to solve the inverse problem, i.e., to describe backward propagation, Kirchhoffmigration was introduced as its adjoint operation that describes the forward propagation of the recorded wavefield in the reverse direction. This turns out to be a very good approximation to backward propagation as long as evanescent waves can be neglected. In other words, whereas the Kirchhoff integral describes a forward extrapolation of the wavefield by summing up all contributions of Huygens secondary sources, the diffraction stack integral describes also a forward extrapolation but backwards in time. However, the underlying principle of constructing the wavefield from elementary Huygens waves is the same.

Up to here, the described methods yield an image which (assuming that the propagation velocity could be estimated sufficiently well) is kinematically correct, i.e., reflectors are mapped to their correct spatial position. It was first realized by Newman (1975) that the amplitude at the image point which results from the summation process could obtain a physically well-defined value if the divergence effects are removed. According to Newman, a "compensation for these effects is mandatory, if reflection amplitudes are to be of diagnostic value". This finally lead to the (often misunderstood) term *true-amplitude* migration. "True-amplitude" in this context means nothing else than the fact that the wavefield is extrapolated in such a way that also dynamic information are used. In other words, the amplitude loss due to geometrical spreading effects of the wavefront is removed, and the resulting amplitude of the reflector image is a measure of the reflection coefficient. Nevertheless, a true-amplitude migration is not able to provide the true reflection coefficient as long as other effects on the amplitude, such as, e.g., source strength, source and receiver coupling effects, transmission loss, attenuation, scattering, or anisotropy, are not corrected for. However, most of these effects are generally slowly varying quantities as a function of source-receiver offset and, therefore, at least a relative measure of the reflection coefficient can be obtained.

2. Theory

Kirchhoff depth migration treats each point M on a sufficiently dense grid in the image domain as a diffraction point. In an a-priori given macrovelocity model the relevant part of the Green's function of a point source at any single diffraction point M in the depth domain is calculated. The kinematic part of this Green's function is the above mentioned configuration-specific Huygens curve/surface. The amplitudes of the input seismograms are stacked along the Huygens curve/surface and assigned to the depth point M. The effect of geometrical spreading is removed from the output amplitudes by multiplying the data during the stack with a true-amplitude weight factor that is calculated from the dynamic part of the Green's function.

Mathematically, the Kirchhoff migration process is expressed as an integration over the recorded wavefield and reads in 3-D (Tygel et al., 1996)

$$V(M) = -\frac{1}{2\pi} \iint_A d\xi_1 d\xi_2 W_{3-D}(\xi, M) \frac{\partial U(\xi, t)}{\partial t}\bigg|_{t=\tau_D(\xi, M)}, \tag{1}$$

where $V(M)$ is the value assigned to one diffraction point M in the depth domain after migration and $U(\xi, t)$ denotes the data in the time domain (seismograms). These data are assumed to consist of analytic traces which allows the handling of complex reflection coefficients (supercritical reflections) and possible caustics along the ray paths. An analytic trace is formed by the actual trace recorded in the field as the real part and its Hilbert transform as the imaginary part. The vector $\xi = (\xi_1, \xi_2)$ is the so-called configuration parameter vector and represents the trace position. Sources and receivers are grouped into pairs, whose locations are described as a function of ξ. The actual form of this function depends on the measurement configuration. The migration aperture A is the area over which ξ varies to cover all source-receiver pairs used in the stack. The factor $W_{3-D}(\xi, M)$ is the true-amplitude weight function. The stacking surface $\tau_D(\xi, M)$ is the above-mentioned Huygens surface. The time derivative is needed in order to correctly recover the source pulse (Newman, 1975).

Assume that the amplitude of one reflection event in the time domain $U(\xi, t)$ is proportional to $R_c \frac{B}{L}$, where R_c denotes the angle-dependent reflectivity, L symbolizes the point-source geometrical spreading factor, and B describes all other effects on the amplitude (some of them were mentioned in the introduction). Then, we want the final true-amplitude signal in the depth domain to be proportional to $R_c B$. This condition leads to a multiplication of the input data during themigration process with a configuration-specific factor, the so-called migration weight function. The mathematical derivation of these weight functions can be found in Tygel et al. (1996) and Jaramillo et al. (1998). A general overview of different weight functions is given in Section 3.

In order to obtain in 3-D the migration result for one point in the output domain, it is necessary to stack amplitudes of the input data along the complete Huygens surface. In the case where model parameters such as the P-wave velocity or the density do not vary in crossline-direction (perpendicular to the acquisition line), the model is True amplitudes: a challenge in reflection seismology called a "2.5-D model". As a consequence, all events in a recorded seismogram stem from reflector elements in the vertical plane through the acquisition line. All data recorded on acquisition lines parallel to the original line would be identical. Since wave propagation in a 2.5-D model is three dimensional, the data contain a full 3-D geometrical spreading effect. In other words, the term 2.5-D corresponds to 3-D wave propagation in a model that has only 2-D parameter variations. The stacking surfaces then shrink to stacking curves. However, the migration operation actually remains a three-dimensional problem to be solved. Fortunately, there is sufficient information to do so as long as the data is really independent from the crossline-direction. In this case, the second integral over ξ_2 of equation (1) can be approximated by means of the method of stationary phase. This procedure leads to the following migration integral for 2.5-D (Martins et al., 1997):

$$
V(M) = \frac{1}{\sqrt{2\pi}} \int_A d\xi W_{2.5-D}(\xi, M) D_t^{-1/2} U(\xi, t)\Big|_{t=\tau_D(\xi, M)} . \tag{2}
$$

The operator $D^{-1\,2}$ denotes the time-reverse half-derivative and is, as in the 3-D case, needed to correctly recover the source pulse.

The weight function in 2.5-D can be seen as a combination of an inline part that merely handles the inline geometrical spreading effect (2-D) and an additional part which handles the out-of-plane geometrical spreading (Bleistein, 1986). Explicit expressions for the weight function can be found in Martins et al. (1997), in particular for the constant velocity case that we have used in the numerical experiments in Section 4.

3. The True-Amplitude Weight Function

In this section, we briefly summarize the weight functions for different cases. In 3-D, the general true-amplitude weight function reads (Jaramillo et al., 1998)

$$
W_{3-D}(\xi, M) = \frac{h_B v_M^2}{2\cos^2(\alpha_M)} L_{SM} L_{MR}, \tag{3}
$$

where v_M is the medium velocity at the image point M and α_M denotes the half-angle between the two ray segments from the source S to the depth point M and from M to the receiver R. The symbols L_{SM} and L_{MR} denote the point-source geometrical spreading factors of these ray segments. The term $h_B = h_B(\xi, M)$ denotes the Beylkin determinant (Beylkin, 1985a,b) given by

$$h_B(\xi, M) = \begin{vmatrix} \nabla \tau_D(\xi, M) \\ \dfrac{\partial}{\partial \xi_1} \nabla \tau_D(\xi, M) \\ \dfrac{\partial}{\partial \xi_2} \nabla \tau_D(\xi, M) \end{vmatrix}. \tag{4}$$

Here $\nabla \tau_D$ symbolizes the gradient of τ_D with respect to the depth coordinates.

For special measurement configurations, the weight function given by equation (3) can be simplified. In the zero-offset case, the true-amplitude migration weight reduces to

$$W_{3-D}^{ZO}(\xi, M) = 4 \frac{\cos \alpha_S}{v_S}(-1)^{\kappa_S} = 4 \frac{\cos \alpha_R}{v_R}(-1)^{\kappa_R}, \tag{5}$$

where α_S and α_R are the ray take-off and emergence angles, and v_S and v_R denote the velocities at the shot and receiver position. The parameters κ_S and κ_R denote the KMAH indices along the ray segments SM and MR, assuming point sources in S and R, respectively. The KMAH index[1] counts the number of caustics along a ray path: it increases by one if the ray tube shrinks to zero in one dimension (caustics of first order), at focusing points (caustics of second order) it increases by two. For the common-shot (CS) case, the weight function reads

$$W_{3-D}^{CS}(\xi, M) = \frac{\cos \alpha_R}{v_R} \frac{|L_{SM}|}{|L_{MR}|} e^{-i\frac{\pi}{2}(\kappa_S + \kappa_R)}. \tag{6}$$

All quantities can be obtained from the dynamic part of the Green's function. The general 2.5-D weight function reads (Martins et al., 1997)

$$W_{2.5-D}(\xi, M) = \frac{v_M^2 \bar{h}_B}{2\cos^2 \alpha_M} \bar{L}_S \bar{L}_R \sqrt{\sigma_{SM} + \sigma_{MR}}, \tag{7}$$

---

[1] The name was chosen to honour the work by Keller, Maslov, Arnold, and Hörmander.

where a bar denotes purely in-plane (2-D) quantities. The out-of-plane effects are described by $\sqrt{\sigma_{SR}}$ where $\sigma_{ab} = \int_a^b v(s)ds$. Here v is the medium velocity and s denotes the arc-length of the ray.

4. Example

In this section, we address the accuracy of true-amplitude migration. This corresponds to the question, "How true are true amplitudes?" Although we have already successfully applied the algorithms to real data sets, we restrict our studies presented in this paper to a synthetic data example. Only in this case, numerical results can be compared to analytical values. Below, we present numerical tests for a simple model consisting of a single dipping layer with a reflector dip of $10°$ and a velocity contrast which leads to a normalincidence reflection coefficient of $R=0.5$. For simplicity, the density was kept constant in both layers. Firstly, a synthetic zero-offset (ZO) section was produced by means of dynamic ray tracing. Secondly, we post-stack migrated the data from the time to the depth domain using a 2.5-D as well as a 3-D algorithm. Finally, the location of the reflector and the amplitudes along the reflector were picked in the migration result. From the theory of true-amplitude migration, we can make the following two statements:
1. Kinematic aspect: The location of the reflector within the migrated image should be identical to the reflector position in the given model.
2. Dynamic aspect: Since there is no transmission loss (or any other process affecting the amplitude except of the geometrical spreading) and the source strength is known, the amplitude along the reflector should be constant and equal to the reflection coefficient $R = 0.5$.

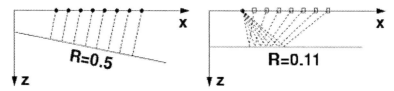

Fig. 2. Sketch of the models used for the numerical examples. Left picture: model for poststack example; a dipping interface with a reflection coefficient of $R = 0.5$. Right picture: model for pre-stack example; a horizontal interface with a zero-offset reflection coefficient of $R = 0.11$.

Figure 3 shows a comparison of the kinematic aspects while Figure 4 compares the amplitudes after migration with the analytical value. The pictures show that both aspects are accurately fulfilled within the limits of discretization. It can be observed that the amplitude error for 2.5-D migration is in this situation smaller than for 3-D migration. This arises because the analytical approximation of the second integral by means of the stationary phase method in 2.5-D schemes is usually better than the calculation by means of a numerical summation used in 3-D (see also Section 2).

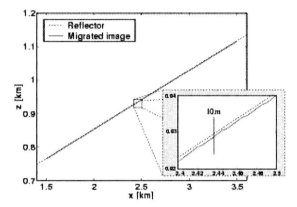

Fig. 3. Comparison of the reflector location after migration and the true location within the model. The spatial grid increments are $\Delta x = \Delta z = 5$m.

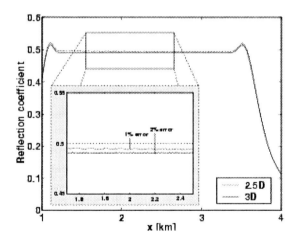

Fig. 4. Comparison of the true reflection coefficient $R = 0.5$ with the picked amplitudes along the reflector image for 2.5-D and 3-D post-stack migration. Sidelobes are due to necessary tapering of the input data because of the finite migration aperture A.

Of course, the true-amplitude algorithms can also be applied to pre-stack data which then allows the extraction of AVO information after migration. To investigate this process, a pre-stack (multicoverage) dataset was created for a situation as depicted in the right part of Figure 2. In this model, a velocity contrast was used that leads to supercritical reflections for offsets greater than 2.6 km. After applying a finite-offset pre-stack depth migration (using the correct velocity model), we obtain a separate migrated image for each offset. If the correct migration velocity was used, these images are kinematically identical. However, the amplitude of the reflector image will be different for different offsets. Consequently, by extracting the amplitude behavior for all offsets, the angle-dependent and, thus, also offset-dependent reflectivity of the target reflector can be retrieved. This is shown in the left part of Figure 5, where the amplitudes picked in the migration result are compared to values that were calculated by means of the Zöppritz equations (Zöppritz, 1919). AVO information obtained in such a way can then be utilized to deduce rock-physical properties of target reflectors such as, e.g., possible hydrocarbon reservoirs. As is well known, the reflectivity becomes complex-valued for supercritical angles. This results in an additional phase shift of the migrated signal for the corresponding offsets. As expected, this behavior can be observed in the right picture of Figure 5, where the phase shift of the migrated signal is compared to the analytical values.

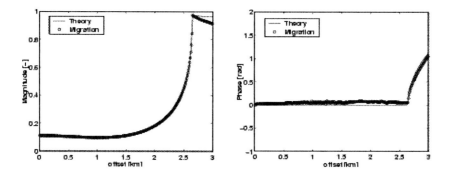

Fig. 5. Amplitude (left part) and phase (right part) of the target reflector for all offsets. Circles denote values that were picked in the migrated image while the solid lines correspond to analytical calculations.

5. Conclusions

In this paper we gave a general overview of Kirchhoff true-amplitude migration: By applying suitable weight functions in the migration process, the geometrical spreading effects - which have a significant influence on the recorded amplitudes - can be removed. The migrated image then provides not only a kinematic image of the subsurface but yields also information about the impedance contrast at interfaces. Trueamplitude migration can thus be seen as a generalization of the classical kinematic migration methods.

Simple numerical experiments have shown that the method is able to retrieve the correct impedance contrast as long as additional influences on the recorded amplitudes may be neglected. In practice, however, true-amplitude migration faces several problems: All migration-preceding processing steps must treat the seismograms in an amplitude-preserving way. In addition, detailed information about amplitudeaffecting medium properties might be necessary. However, true-amplitudemigration is in most cases able to allow at least a relative comparison of reflection impedances.

There exist several approximations of the general weight functions presented in this paper. These can be applied without much additional computational effort compared to a purely kinematic migration and are, therefore, suited for practical use in the exploration industry. The true-amplitude theory is still under development and will further improve in future. For example, true-amplitude migration of converted waves and the involvement of anisotropy is topic of current research.

Acknowledgements

This work was kindly supported by the sponsors of the Wave Inversion Technology (WIT) Consortium, Karlsruhe, Germany. Moreover, we are grateful to Prof. Karl Fuchs that we were allowed to present our scientific work at the symposium "Challenges for earth sciences in the 21st century".

References

Beylkin, G. (1985a). Imaging of discontinuities in the scattering problem by inversion of a generalized Radon transform. J. Math. Phys., 26:99–108.

Beylkin, G. (1985b). Reconstructing discontinuities in multidimensional inverse scattering problems. Applied Optics, 24:4086–4088.

Bleistein, N. (1986). Two-and-one-half dimensional in-plane wave propagation. Geophysical Prospecting, 34:686–703.

Bleistein, N. (1999). Hagedoorn told us how to do Kirchhoff migration and inversion. The Leading Edge, 18(8):918–927.

Hagedoorn, J. (1954). A process of seismic reflection interpretation. Geophysical Prospecting, 2(2):85–127.

Jaramillo, H., Schleicher, J., and Tygel, M. (1998). Discussion and Errata to: A unified approach to 3-D seismic reflection imaging, Part II: Theory. Geophysics, 63(2):670–673.

Martins, J., Schleicher, J., Tygel, M., and Santos, L. (1997). 2.5-D True-amplitude Migration and Demigration. J. Seis. Expl., 6(2/3):159–180.

Newman, P. (1973). Divergence Effects in a Layered Earth. Geophysics, 38(3):481–488.

Newman, P. (1975) Amplitude and phase properties of a digital migration process. In: 37th Ann. Internat. Mtg., Europ. Assoc. Expl. Geoph., Expanded Abstracts, Bergen. (Republished in: First Break, 8, 397–403, 1990).

Schneider, W. (1978). Integral formulation for migration in two and three dimensions. Geo-physics, 43(1):49–76.

Tygel, M., Schleicher, J., and Hubral, P. (1995). Dualities between reflectors and reflection-time surfaces. J. Seis. Expl., 4(2):123–150.

Tygel, M., Schleicher, J., and Hubral, P. (1996). A unified approach to 3-D seismic reflection imaging, Part II: Theory. Geophysics, 61(3):759–775.

Zöppritz, K. (1919). Erdbebenwellen VIII b: Über Reflexionen und Durchgang seismischer Wellen durch Unstetigkeitsflächen. Göttinger Nachrichten, 1:66–84.

Imaging of Orebodies with Vertical Seismic Profiling Data

Eric Duveneck[1], Christof Müller[2], and Thomas Bohlen[2]

[1] Geophysical Institute, University of Karlsruhe, Germany
[2] Institute of Geosciences, Kiel University, Germany

Abstract

Geophysical exploration for volcanogenic massive sulphide (VMS) deposits, a major source for lead, copper, zinc and other metals, is conventionally done using magnetic and electromagnetic methods. For small deposits at depths greater than about 500 m, where these methods are no longer applicable, vertical seismic profiling (VSP) methods are a potential alternative, when a borehole exists in the area to be examined.

There are, however, a number of problems that make it difficult to uniquely locate orebodies in the 3D subsurface from VSP data, especially when only a limited number of source positions at the surface is used. Among these problems are spatial ambiguity, the complexity of the scattered elastic wavefield, and the presence of subsurface reflectors apart from the orebody itself.

A method is presented that makes use of polarisation information present in multicomponent VSP data to tackle some of these problems and directly locate isolated scattering objects of high impedance contrast in the vicinity of the borehole. This *Polarisation Migration* is successfully tested on synthetic elastic Finite-Difference (FD) seismograms of the pure scattering response of a complex object in a simple VSP geometry.

In order to apply the method to field data a number of requirements must be met in preprocessing: Relative amplitudes of the three components must be preserved while the direct wavefield must be completely removed from the data. A VSP dataset measured near a known orebody is processed and migrated using the described Polarisation Migration. The migration result shows a clear anomaly at the actual position of the orebody.

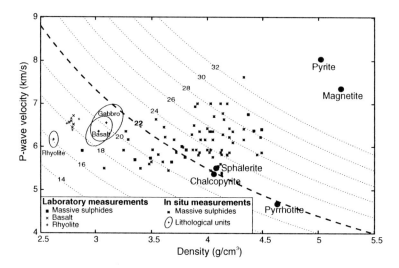

Fig. 1. Laboratory and in situ measurements of P-wave velocities versus densities for sulphide minerals and typical surrounding rocks. Lines of constant seismic impedance are also drawn. Pyrite is characterized by a very high seismic impedance (from: Adam et al. 1997).

1. Introduction

Volcanogenic massive sulphide (VMS) deposits play a major role in the exploration for a number of economically important metals like lead, copper and zinc. These deposits usually occur as relatively small (a few hundred meters extension) bodies of submarine-volcanic origin. While traditional geophysical exploration for ore deposits relies heavily on magnetic and electromagnetic methods, these methods are of limited use in the exploration for small, deep deposits (Adam 2000).

Recently, a number of investigations have been carried out in order to test the applicability of seismic methods in the exploration for VMS deposits. Results of Salisbury et al. (1999) show that the ore mineral composition of such deposits, especially due to the high content of pyrite, is characterized by a very high density contrast compared to typical volcanic surrounding rocks (Fig. 1). They should consequently be visible in seismic reflection data.

A number of 2D and 3D seismic reflection experiments have been carried out during the past years in the area of known VMS deposits to verify this statement and to investigate the specific challenges encountered in crystalline environments. Some of these experiments are documented in Adam et al. (1997, 1998) and Calvert and Li (1999).

A major disadvantage of surface seismic reflection measurements for mineral exploration is, however, the high costs of such experiments. As an alternative, if boreholes exist (e.g. in areas of ongoing mining), these can be used for carrying out relatively cheap seismic downhole experiments to locate new deposits in the vicinity of the boreholes. Again, the penetration depth of seismic waves should be superior to that of downhole electromagnetic methods. A possible measurement geometry for such vertical seismic profiling (VSP) experiments could include a string of geophones in the borehole and a number of sources at varying distances and azimuths around the borehole at the surface.

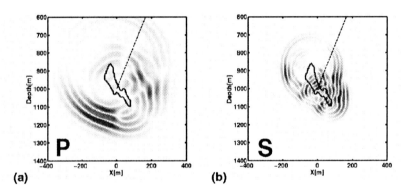

(a)   (b)

Fig. 2. Snapshots of the pure scattering response of an orebody model due to a plane P-wave of 50 Hz dominant frequency, modelled with 3D elastic Finite Differences (FD). The incidence direction is indicated by a black dashed line. See text for details. **(a)** Scattered P- **(b)** scattered S-wavefield.

The purpose of these VSP measurements, i.e. the direct location of small isolated orebodies, deviates significantly from the conventional use of VSP measurements as an interpretational aid for surface reflection seismic data in sedimentary environments (e.g. Hardage 2000). Consequently, new specific problems arise in the processing and interpretation of the resulting data, which have not been addressed before. These are mainly caused by the sparseness of the measurement geometry and the complexity of the wavefield scattered by an orebody of possibly complex shape.

Here, we present a simple method that makes use of three-component seismic data recorded in a borehole to directly locate small objects of high impedance contrast in the vicinity of the borehole. A similar method has also been proposed by Takahashi (1995).

2. Description of the Problem

The problem treated here can be stated as follows: locate an isolated object of complex shape and high density contrast that is situated somewhere in the vicinity of a borehole. Do this by means of seismic measurements involving an array of receivers (geophones) in the borehole and a limited number of source positions at the surface around the borehole. The borehole depth, as well as the depth of the object to be located, is assumed to be of the order of 1 km.

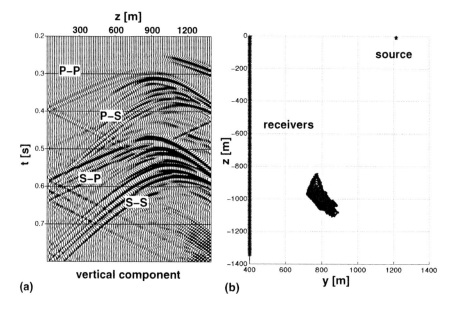

(a)   (b)

Fig. 3. Elastic FD simulation of a simple offset VSP experiment. (a) vertical component of the pure scattering response from a model orebody. (b) measurement geometry. A number of interfering scattering modes are visible in the seismogram, together with reflections from the free surface and some modelling artefacts.

2.1 Scattering of the Elastic Wavefield

The objects of interest are orebodies of about 100 to 500 m length and are characterized by a very high density contrast ($\delta\rho/\rho \approx 0.5$) compared to typical volcanic surrounding rocks. They are typically of a relatively complex shape with rough edges. Their size and shape imply that for common frequency contents of seismic sources these objects act as scatterers. The scattered wavefield is expected to consist of a number of interfering scattering modes including P-P and S-S scattering, as well as P-S and S-P conversions.

Figures 2a and 2b show snapshots of the modelled pure scattered P- and S-wavefield due to an incident plane P-wave of 50 Hz dominant frequency. These were calculated with 3D elastic Finite Differences (FD) (Bohlen 2002), using the shape of a known VMS deposit as a model. The direction of the incident wavefield is given by the dashed straight line in each figure. The shape of the scatterer in the displayed vertical plane is also indicated in black. Model parameters were v_p=6000 m/s, v_s=3000 m/s, ρ=2.7 g/cm^3 for the background and v_p=5500 m/s, v_s=2680 m/s, ρ=4.08 g/cm^3 for the orebody, which corresponds to Sphalerite (see Fig. 1). The maximum energy in the P-wavefield snapshot in Fig. 2a, normalised to the incident P-wave energy, is E_{max}=0.302, while the maximum energy of the S-wavefield in Fig. 2b normalised to the incident P-wave energy is E_{max}=0.280.

It can be observed that the scattered wavefield displays considerable complexity with strongly varying angle-dependent amplitudes. This amplitude variation is due not only to the shape of the scattering object (e.g. focussing effects), but also depends on its composition in contrast to the surrounding material (Bohlen et al. 2002).

Because of the high density contrast and relatively large dimensions of VMS orebodies compared to typical seismic wavelengths in exploration, the scattered wavefield cannot be described by the Born approximation for elastic wave scattering (Gubernatis et al. 1977; Müller 2000). Consequently the influence of shape and composition of the scattering object on the amplitudes of the scattered wavefield cannot be separated and no simple description of the scattered wavefield is possible (Bohlen et al. 2002). It can be shown, though (e.g., Müller 2000), that even for the case of a spherical homogeneous scatterer, the seismic impedance (the product of velocity and density) contrast does not give a useful measure of scattering strength, as it would for reflections from a planar interface. Even with no impedance contrast at all, scattering may be strong, if velocity and density contrasts are present.

Figure 3a shows the pure scattering response of the same model orebody that would be measured on the vertical (z) component in a simple VSP experiment as illustrated in Fig. 3b (the direct P- and S-waves have been removed by subtraction). A complex wavefield, consisting of a number of different interfering scattering modes can be seen. Due to different traveltimes and moveouts P-P, P-S, S-P and S-S phases are discernible. Again strong angle-dependent amplitude variations can be observed, that also vary significantly between different scattering modes.

This strong interference of different scattering modes showing different angle-dependent amplitude variations can lead to serious problems for imaging of orebodies using the measured elastic wavefield. In general, seismic imaging (migration) algorithms are designed to handle one wave type (converted or non-converted) only. Interfering scattering modes like those seen in Fig. 2a can cause severe imaging artefacts, that, depending on the amplitudes, can completely mask the desired image of the orebody.

2.2 Spatial Ambiguity

Another problem that arises due to the sparse measurement geometry is spatial ambiguity. Even in two dimensions the traveltimes recorded in a borehole do not allow to uniquely deduce the true subsurface position of a scattering object, when only a single shot position at the surface is used. The scattering object could be located on either side of the borehole.

To locate an object in the three-dimensional subsurface using only the recorded traveltimes from one surface source position is impossible. The problem is azimuthally completely ambiguous. The situation improves to a certain degree if a number of source positions at different azimuths around the borehole are used, but as long as there is no complete source coverage of the surface area around the borehole, it will be difficult to image a scattering object with conventional seismic imaging methods (Müller 2000).

The reason for this is the fact that seismic imaging (migration) methods are usually based on extrapolation of the recorded wavefield into the subsurface and application of an imaging condition (Claerbout 1985; Chang and McMechan 1986). The wavefield recorded by an array of receivers in a borehole due to a given shot at the surface cannot be extrapolated into the subsurface unless additional information on the azimuthal direction of wave propagation is available. The signal could have reached the receiver from any azimuthal direction.

If a dense 2D array of sources at the surface around the borehole had been used, then the signals recorded at each receiver due to all sources could be used for imaging; assuming reciprocity these signals could be viewed as the wavefield recorded areally at the surface due to a source at depth and could be extrapolated into the subsurface.

Taken together, the complexity of the scattered wavefield, consisting of a number of interfering scattering modes, and the spatial ambiguity caused by a sparse measurement geometry make it difficult to image orebodies in the three-dimensional subsurface using scalar VSP measurements. A possible key to the solution of both of the described difficulties lies in the use of additional wavefield information that is available, if the seismic wavefield has been recorded with three-component geophones: polarisation.

3. Using Polarisation Information

For simplicity, we initially consider the case of Rayleigh scattering, where kR<<1, i.e. the characteristic length R of the scattering object is much shorter than the wavelength $\lambda=2\pi/k$ of the incident wavefield. For this case the polarisation of the scattered wavefield in the far-field is independent of the contrast in density and the Lamé elastic parameters ($\delta\rho,\delta\lambda,\delta\mu$) for P-P, P-S and S-P scattering (Korneev and Johnson 1996). This result still approximately holds for homogeneous spheres of a size comparable to the dominant wavelength of the incident wavefield. For S-S scattering, though, the polarisation of the scattered wavefield does depend on the contrasts in material parameters of the scattering spherical object. These results can be written in a coordinate-system-independent form as (Wu 1989):

$$\mathbf{u}_{pp} = c_{pp} \cdot \hat{\mathbf{o}} \qquad (1)$$

$$\mathbf{u}_{ps} = c_{ps} \cdot (\hat{\mathbf{a}} - \hat{\mathbf{o}}(\hat{\mathbf{o}} \cdot \hat{\mathbf{a}}))$$

$$\mathbf{u}_{sp} = c_{sp} \cdot \hat{\mathbf{o}}$$

$$\mathbf{u}_{ss} = c_{ss}^{I} \cdot (\hat{\mathbf{a}} - \hat{\mathbf{o}}(\hat{\mathbf{o}} \cdot \hat{\mathbf{a}})) + c_{ss}^{II} \cdot (\hat{\mathbf{i}} - \hat{\mathbf{o}}(\hat{\mathbf{o}} \cdot \hat{\mathbf{i}}))$$

where $\mathbf{u}$ are displacement vectors, the subscripts denote the scattering mode (e.g. $\mathbf{u}_{ps}$ for an incident P-wave and a scattered S-wave), and the coefficients c are dependent on the material parameters inside and outside of the scattering object.

They are proportional to the squared frequency of the incident wavefield and are strongly angle-dependent. The vector $\hat{\mathbf{a}}$ is the polarisation of the incident wavefield, $\hat{\mathbf{i}}$ is the normalised slowness vector, giving the ray direction of the incident wavefield, and $\hat{\mathbf{o}}$ is the normalised slowness vector (ray direction) of the scattered wavefield.

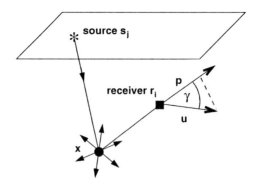

Fig. 4. For a given source-receiver combination, a given subsurface point **x** and a given scattering mode (in this example P-P) the angle γ between the measure polarisation (**u**/|**u**|) and the expected polarisation **p** can be determined. The assumption made is that there is a Rayleigh scatterer at **x**.

This characteristic polarisation of P-P, P-S and S-P scattered wavefields can be used to identify and locate isolated Rayleigh or spherical scatterers in the three-dimensional subsurface from measured multicomponent seismograms. It can be built into an imaging algorithm as follows (Duveneck et al. 2001):

For a given subsurface point **x** the expected polarisation **p** at every receiver position can be calculated for any scattering mode if a Rayleigh scatterer is assumed at **x**. The angle γ between **p** and the wavefield **u** that was actually measured at the corresponding traveltime (see Fig. 4) can be determined ($\cos(\gamma)=\mathbf{p}\cdot\mathbf{u}/|\mathbf{u}|$) and used in a diffraction stack type migration, where the sum along diffraction traveltime curves in the data is not carried out over amplitudes, but over a function $f(\gamma)$ yet to be defined. The value of the migrated image at the subsurface point **x** is then given by:

$$P(\mathbf{x}) = \frac{1}{MN} \sum_{i=1}^{M} \sum_{j=1}^{N} f(\gamma(\mathbf{r}_i, \mathbf{x}, \mathbf{s}_j)), \tag{2}$$

where M receivers $\mathbf{r}_i$ and N sources $\mathbf{s}_j$ are used (see Fig. 4). Here, $\gamma(\mathbf{r}_i, \mathbf{x}, \mathbf{s}_j)= \gamma(\mathbf{u}(\mathbf{r}_i, \mathbf{s}_j, \tau(\mathbf{r}_i, \mathbf{x}, \mathbf{s}_j)), \mathbf{p}(\mathbf{r}_i, \mathbf{x}, \mathbf{s}_j))$. The polarisation **p** and the traveltime τ depend on the considered scattering mode. Assuming ,e.g., constant velocities and P-S scattering, $\mathbf{p}=\mathbf{u}_{ps}/|\mathbf{u}_{ps}|$ (compare Eq. (1)), and $\tau=|\mathbf{x}-\mathbf{s}_j| / v_p + |\mathbf{r}_i-\mathbf{x}| / v_s$.

If high values of $|P(\mathbf{x})|$ are to be interpreted as indicating the presence of a Rayleigh scatterer at **x**, the function $f(\gamma)$ needs to be defined accordingly. It should be small for high values of γ, suppressing data points that are unlikely to originate from point scatterers. f(γ) may or may not explicitly depend on the measured amplitude |u|.

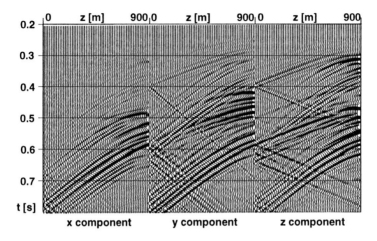

Fig. 5. 3D elastic FD synthetic seismograms calculated for the VSP geometry in Fig. 3b. Every third trace of the pure scattering response (no direct waves) is shown, amplitudes are clipped. See text for details.

In order to apply the described method to the detection of orebodies (which are much too large to be considered as Rayleigh scatterers and far from spherical in shape) assumptions need to be made about scattering from these structures. VMS deposits are in general of a complicated shape with many sharp edges. If one assumes that scattering from these edges shows the same characteristic polarisation as that of Rayleigh scattering, ore deposits can be detected by imaging these edges. This assumption may seem inadequate, but as the following synthetic example shows, it leads to useful results.

4. A Synthetic Data Example

In order to test the proposed imaging approach it was applied to the simple synthetic data example already introduced in Fig. 3, which has been produced by 3D elastic FD modelling. The three-dimensional shape of a known canadian orebody, a typical VMS deposit, was used in the modelling. The homogeneous background material parameters were v_p=6000 m/s, v_s=3000 m/s, ρ=2.7 g/cm^3. The orebody material parameters were v_p=5500 m/s, v_s=2750 m/s, ρ=4.3 g/cm^3. This results in a contrast of the elastic moduli of $\delta\lambda/\lambda=\delta\mu/\mu =0.34$ and of the density of $\delta\rho/\rho=0.59$. These are close to the expected values for a real VMS deposit (Salisbury et al. 1999).

The three displacement components were simulated at 270 receiver positions with a vertical receiver separation of 5 m. The direct wavefield was calculated for a homogeneous model and subtracted from the total wavefield, leaving only the pure scattering response. Figure 5 shows every third trace of the portion of the three-component seismograms that were actually used for imaging.

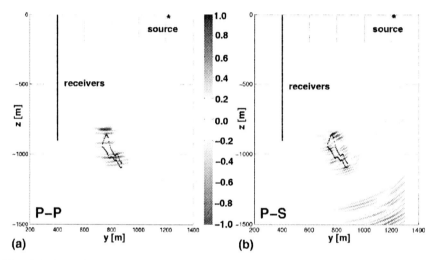

Fig. 6. Polarisation Migration results for synthetic three-component VSP data (see Fig. 5). **(a)** P-P migration result using f_I of Eqs. (3) **(b)** P-S migration result using f_{II} of Eqs. (3). The noise below 1200 m depth is caused by S-S scattered waves in the data. The outline of the model orebody is indicated by black dots.

The downgoing part of the scattered wavefield (forward scattering) was discarded as it could not be separated from the downgoing direct wavefield in practice. The Polarisation Migration as described above was applied to these data.

Due to the strong angle-dependence of scattered wave amplitudes and great differences in amplitude for different interfering scattering modes it proved necessary to suppress the measured amplitude values $|\mathbf{u}|$ and let f depend on γ only. The following two functions $f(\gamma)$ were used:

$$f_I(\gamma) = \begin{cases} 1 & \text{if } |\gamma| < 6° \\ -1 & \text{if } 180° - |\gamma| < 6° \\ 0 & \text{else} \end{cases} \tag{3}$$

$$f_{II}(\gamma) = \cos^9(\gamma)$$

That way, data values that represent energy scattered by a Rayleigh scatterer are automatically extracted, even if surrounded by events of higher amplitude. In this respect the method described here differs from that of Takahashi (1995), who proposes a similar approach, but uses the polarisation information in a weighting factor for the measured amplitude, considering only P-P scattering. His method is based more directly on the Generalized Radon Transform (GRT) migration of Miller et al. (1987), which applies additional weights and is formulated with the assumption of acoustic Born scattering.

Vertical slices through the migration results for P-P and P-S scattering are diplayed in Fig. 6. For the P-P case (a) f_I of Eqs. (3) was used, while for the P-S case(b) f_{II} was used. The position of the orebody within the displayed slices is indicated by black dots. In both cases the true position of the orebody is characterized by high migration values $|P(x)|$.

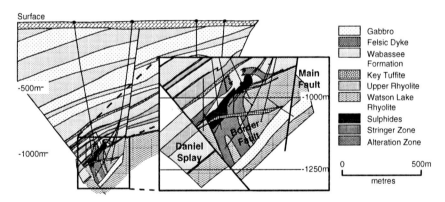

Fig. 7. Geological setting of the Bell Allard VMS deposit (black) (from: Adam et al. 1997)

Even though only one source position was used, the spatial location of the scattering object could be uniquely determined within the vertical plane containing source and receivers. Spatial resolution in the direction normal to this plane is slightly worse (depending on whether f_I or f_{II} has been used), as the spatial ambiguity associated with the recording geometry is higher in this direction. In Fig. 6b some noise is visible at greater depths. This is due to high amplitude S-S scattering modes present in the data, which could not be completely suppressed by the migration, as their polarisation is very similar to that of the P-S modes. A closer look at the P-S migration result in Fig. 6b gives a justification of the assumptions made earlier about scattering from the edges of objects of complex shape; high migration values are visible at different prominent parts of the orebody.

5. A Real Data Example

For the application of the described Polarisation Migration method to real data a number of conditions need to be met: the orientation and relative amplitudes of the different geophone components need to be known. During the entire processing relative amplitudes of the three components need to be preserved and the downgoing wavefield needs to be completely removed without affecting the polarisation of the upgoing, scattered waves. This last requirement demands a high signal-to-noise ratio.

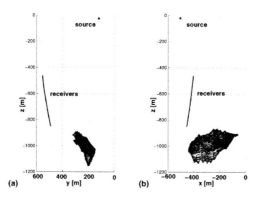

Fig. 8. Acquisition geometry of the VSP experiment near Bell Allard orebody. **(a)** looking east **(b)** looking north. Only receivers situated in a small portion of the borehole could be used for imaging.

The Polarisation Migration has been applied to seismic data recorded in a borehole close to a known VMS deposit, the Bell Allard orebody in Matagami mining camp, Québec, Canada (Adam et al. 1998; Adam 2000). The orebody is located at about 1000 m depth and was discovered in 1992. It is 370 m long and about 165 m wide, and its estimated total mass is 6,000,000 t. A vertical geological cross-section from north to south through the orebody and the surrounding geological units is given in Fig. 7. What makes this orebody suitable for a test of the described imaging method is the fact that its precise shape and composition are well known from drilling and mining and that VSP data have been recorded in its direct vicinity. The VSP data used here were recorded in a borehole about 200 m away from the orebody (Adam 2000). For the experiment a single source position about 400 m south of the wellhead and three-component geophones at depths between 113 m and 850 m were utilised. Of these, only receiver locations between 465 m and 850 m with a depth interval of 5 m could be used for imaging (Fig. 8). This very sparse dataset is far from ideal for the task of imaging orebodies. Apart from the fact that only one source position was used, the traces are severely contaminated by electrical noise and reverberations that could not be completely removed. The overall signal-to-noise ratio is relatively low.

The following processing steps were applied after an initial removal of electrical noise at 60 Hz and its multiples (Adam 2000):

- Removal of bad traces (trace editing)
- Bandpass filter (20, 40, 160, 200 Hz)
- Static corrections (based on direct P-wave)
- Determination of the orientation of geophone components (based on direct P-wave amplitudes)
- Rotation of the components into a coordinate system with the x-axis oriented towards east
- Removal of the downgoing wavefield by application of an f-k-filter

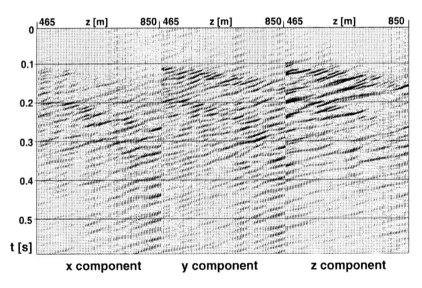

Fig. 9. The three components of the upgoing wavefield. The direct waves have been removed by application of an f-k-filter.

The f-k-filtered seismograms (Fig. 9) served as the input to the Polarisation Migration. The P- and S-wave velocities could be directly read from the slope of the direct waves and appear to be constant over the entire depth range. They are assumed to be isotropic and their values are v_p=6378 m/s and v_s=3591 m/s. The Polarisation Migration was applied for P-P scattering with $f_l(\gamma)$ of Eqs. (3). The results are displayed in Fig. 10 as a depth slice at 1025 m depth (a) and as an iso- surface plot at the value of –0.4 of the normalisaed migration volume (b). A clear migration response can be seen at the true position of the orebody, though not at the expected top edge of it. An explanation of this could lie in the fact that the orebody is not a homogeneous body: Adam et al. (1997) observe that in surface seismic reflection data the lower part of the orebody appears to be much more reflective than the upper part. They attribute this to the fact that the lower part is pyrite- and magnetite-rich and therefore has a much higher seismic impedance than the upper, sphalerite-rich part of the deposit. Also, the migration response could be due to the alteration zone directly beneath the orebody itself, which is also strongly enriched in pyrite.

6. Conclusions

We have described a new simple method of locating isolated objects of high contrast (e.g. VMS deposits) in the direct vicinity of boreholes using the polarisation information present in multicomponent VSP recordings. By summing over a very restrictive function of polarisation fit during migration, signals that are likely to originate from isolated scattering objects are emphasized over other signals. The method has successfully been tested on a synthetic example and was applied to a single-source VSP dataset recorded near a known VMS deposit.

In contrast to conventional migration methods, the spatial resolution of the method described here depends not only on the wavenumber coverage at each subsurface location, associated with the given acquisition geometry, but mainly on the choice of the polarisation fit function. If the polarisation information in the data is not degraded by noise, a very restrictive function may be used, resulting in an improved spatial resolution. Otherwise a less restrictive polarisation fit function has to be used.

The influence of noise or multiple scattering in a heterogeneous background medium on the imaging results still needs to be investigated, as does the effect of anisotropy.

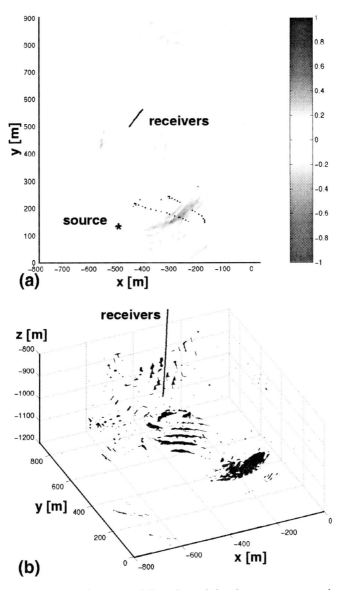

Fig. 10. Result of P-P Polarisation Migration of the three-component data in Fig. 10, using $f_{II}(\gamma)$ in Eqs. (3). **(a)** Depth slice through the normalised migrated volume at 1025 m depth. High migration amplitudes are visible at the actual location of the orebody. An outline of the orebody is indicated by black dots. **(b)** Isosurface plot of the normalised migration volume at a value of –0.4

Acknowledgement

This work was carried out in cooperation with the Downhole Seismic Imaging Consortium for Mineral Exploration (DSI). The BAS-92-25 VSP dataset was acquired by Doug Schmitt (University of Alberta).

References

Adam E (2000) Détection directe d'ungisement de sulfures massifs volcanogénes par sismique réflexion (in English). Ph.D. thesis, École Polytechnique de Montréal

Adam E, Arnold G, Beaudry C, Matthews L, Milkereit B, Perron G, Pineault R (1997) Seismic exploration for VMS deposits, Matagami, Québec. In: Gubins A G (ed) Proceedings of Exploration 97: Fourth Decennial International Conference on Mineral Exploration

Adam E, Milkereit B, Mareschal M (1998) Seismic reflection and boreholegeophysical investigations in the Matagami mining camp. Can. J. Earth Sci. 35:686-695

Bohlen T (2002) Parallel 3-D viscoelastic finite-difference modelling. Computers & Geosciences, 28 (8): 887-899

Bohlen T, Müller C, Milkereit B (2002) Elastic wave scattering from massive sulphide orebodies: on the role of composition and shape. To appear in: Hardrock Seismic Exploration, SEG Developments in Geophysics

Calvert A J, Li Y (1999) Seismic reflection imaging over a massive sufide deposit at the Matagami mining camp, Québec . Geophysics 64: 24-32

Chang W F, McMechan G A (1986) Reverse-time migration of offset vertical seismic profiling data using the excitation-time imaging condition. Geophysics 51: 67-84

Claerbout J F (1985) Imaging the earths interior. Blackwell Science Publishing

Duveneck E, Müller C, Bohlen T (2001) Imaging of orebodies with a Polarization Stack Migration. In: 63rd EAGE meeting, Amsterdam, Netherlands, Session M-14

Gubernatis J E, Domany E, Krumhansl J A, Huberman M (1977) The Born approximation in the theory of the scattering of elastic waves by flaws. J. Appl. Phys. 48: 2812-2819

Hardage B A (2000) Vertical Seismic Profiling: Principles. 3rd edn Handbook of Geophysical Exploration, Seismic Exploration , vol 14A, Pergamon

Korneev V A, Johnson L R (1996) Scattering of P and S waves by a sperically symmetric inclusion. Pure and Appl. Geophys. 147: 675-718

Miller D, Oristaglio M, Beylkin G (1987) A new slant on seismic imaging: migration and integral geometry. Geophysics 52: 943-964

Müller C (2000) On the nature of scattering from isolated perturbations in elastic media and the consequences for processing of seismic data. Ph.D. thesis, Christian-Albrechts-Universität zu Kiel

Salisbury M H, Bleeker W, Eaton D, Milkereit B (1999) Detection of massive sulfides at
Kidd Creek using logging, rock properties, and seismic reflection methods. Economic Geology 10: 541-554

Takahashi T (1995) Prestack migration using arrival angle information. Geophysics 60: 154-163

Wu R-S (1989) The perturbation method in elastic wave scattering. Pure and Appl. Geophys. 131: 605-637

The Dead Sea Fault and Its Effect on Civilization

Zvi Ben-Avraham, Michael Lazar, Uri Schattner, and Shmuel Marco

Department of Geophysics and Planetary Sciences, Tel Aviv University, Israel

Abstract

The Dead Sea fault (DSF) is the most impressive tectonic feature in the Middle East. It is a plate boundary, which transfers sea floor spreading in the Red Sea to the Taurus collision zone in Turkey. The DSF has influenced many aspects of this region, including seismicity and ground water availability. It may have even affected the course of human evolution.

Numerous geophysical and geological studies of the Dead Sea fault provide insight into its structure and evolution. Crustal structure studies have shown that the crust at the fault zone is slightly thinner than that of the regions west and east of it. A transition zone between the lower crust and the Moho under the fault was mapped.

The region has a remarkable paleoseismic record going back to about 70 ka years. Several earthquakes, such as the one that occurred in the Dead Sea region on 31 BC, may have even influenced the course of history of this region. The confusion and fear inflicted by the earthquake paved the way for the expansion of Herod's kingdom. Places such as Jericho, the oldest city in the world, which are located within the valley formed by the fault, were affected immensely by seismic activity.

The DSF is an important part of the corridor through which hominids set off out of Africa. Remains of the earliest hominids are found in several sites along the Dead Sea fault, including Erk-el-Ahmar, Ubediya and Gesher Benot Ya'aqov. It is interesting to note that acceleration in the vertical motion along the Dead Sea fault, which produced its present physiography, began slightly before man had started his way out of Africa northwards.

1. Introduction

The Dead Sea fault (DSF) is the most impressive tectonic feature in the Middle East (Fig. 1). It is a left lateral transform plate boundary, separating the Arabian plate and the Sinai sub-plate. The transform has been active since the Miocene (Garfunkel, 1981; Garfunkel and Ben-Avraham, 1996) with movement continuing today. Motion is transferred along the fault from the opening of the Red Sea in the south, to the Taurus-Zagros collision zone in Turkey and Iran to the north. It is one of the most seismically active regions in the Middle East.

The region has a remarkable historical and geological record of seismicity going back to about 70 ka. Several historical earthquakes have caused extensive damage in the area. Places such as Jericho, the oldest city in the world, or Bet She'an (Fig. 2), one of the largest cities in the region in Roman time, were greatly affected by seismic activity.

Crustal structure studies have shown that the crust directly under the fault valley is somewhat different from that on the sides. These differences in crustal structure may have controlled the evolution of physiography in the region. Since the Miocene the margins of the transform were uplifted in several stages and the rift floor subsided, creating the present-day physiography. A large part of the transform is situated below sea level. The lowest place along the transform (and on Earth) is the Dead Sea basin (Ben-Avraham, 2001).

The physiography of the Dead Sea fault is a result of vertical motion, which caused subsidence of the floor of the rift and uplift of its shoulders. Acceleration in the vertical motion began shortly before man started his way out of Africa. Remains of the most ancient hominids outside Africa are found along the DSF, which actually formed a corridor through which hominids set off out of Africa. The geological evolution of the DSF and the active tectonic processes occurring along its length, thus, may have affected the course of human history.

2. Crustal Structure

"In his hand are the deep places of the earth: and the strength of the hills is his also. The sea is his and he made it: and his hands formed the dry land" Psalm 95:4-5

Seismic refraction studies along the DSF indicate that the crust directly under the fault valley is thinner than that on both sides. It is 33 km thick at

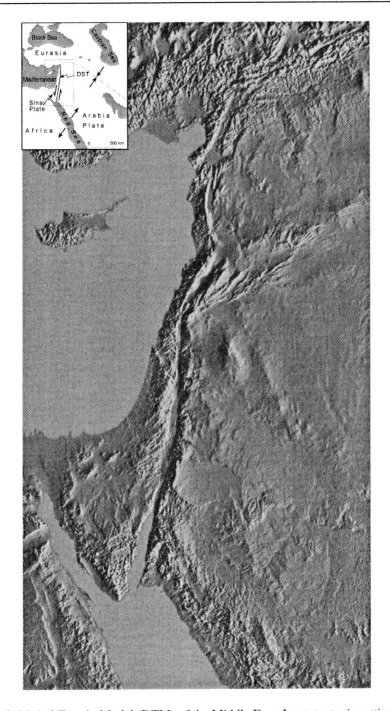

Fig. 1. Digital Terrain Model (DTM) of the Middle East. Inset: tectonic setting

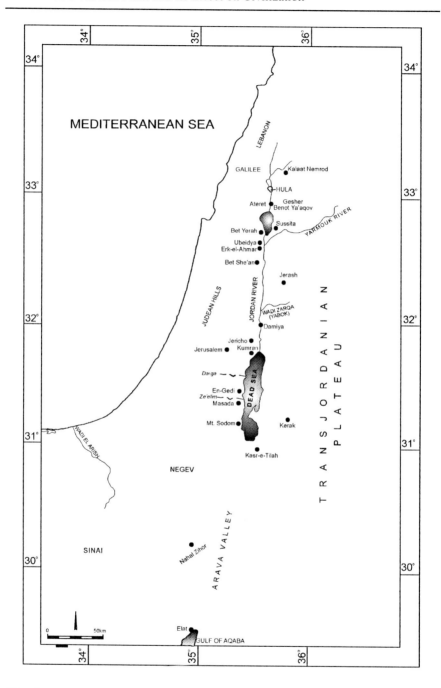

Fig. 2. Sites along the Dead Sea fault mentioned in the text (modified after Horowitz, 2001)

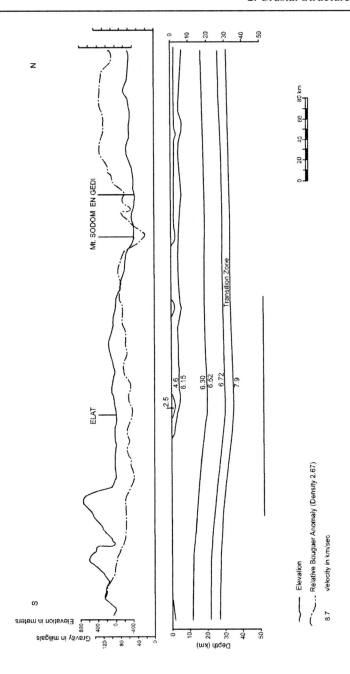

Fig. 3. A composite section along the Dead Sea Fault from the Sea of Galilee to the Red Sea showing the calculated crustal model and the relative Bouguer anomaly (after Ginzburg, 1981). The locations of Elat, Mt. Sodom, and En Gedi are shown in Fig. 2

the Dead Sea basin and shows a slight thickening towards Elat. From Elat south the crust thins gradually from a thickness of 35 km to a thickness of 27 km, 160 km south of Elat (Ginzburg et al., 1979). A discontinuity between the upper and lower crust was also observed (Fig. 3). The seismic refraction data also show a 5 km thick velocity transition zone within the lower crust above the crust-mantle boundary. In other areas adjacent to the fault the crust-mantle boundary is manifested as a sharp velocity discontinuity. However, many questions still remain as to the detailed crustal structure along the DSF (DESERT group, 2000).

The Dead Sea basin itself provides evidence for dramatic activity during the Plio-Pleistocene. Crustal studies indicate that a thick sedimentary fill characterizes the two basins of the Dead Sea graben. Depth to basement is about 6 km in the northern basin and more than 12 km in the southern basin (Fig. 4) (Ginzburg and Ben-Avraham, 1997). Local earthquake data indicate the presence of lower crustal seismicity under the transform. In the Dead Sea basin, hypocenters are located almost as deep as the Moho discontinuity (Aldersons, in preparation).

The changes in crustal thickness across the transform implies that it is a relatively narrow zone of deformation, which penetrates the entire crust (ten Brink et al., 1990). This unique crustal structure allowed the dramatic uplift of the transform margins and subsidence of the floor of the fault since the Miocene until present.

3. Evolution of Physiography

"Every valley shall be exalted, and every mountain and hill shall be made low; and the crooked shall be made straight, and the rough places plain" Isaiah 40:4.

The prominent morphotectonic expression of the DSF (Fig. 1) is characterized by the deepest continental depressions in the world, flanked by up to ~3-km-high margins (Fig. 1). In general, the eastern shoulder rises gently towards the west, reaching its highest elevations near the transform and drops abruptly into the median valley. The elevation of the eastern shoulder is usually higher than that of the western shoulder, reflecting broad regional uplift along the rift (Wdowinski and Zilberman, 1997).

Present-day physiography of the DSF formed as a result of continental breakup processes took place since the beginning of the Miocene throughout the Present. In the Plio-Pleistocene, a major tectonic phase uplifted the margins. During this period the Arava rift valley subsided and the Negev was uplifted and tilted (Avni, 1998). The accelerated uplift at the end of

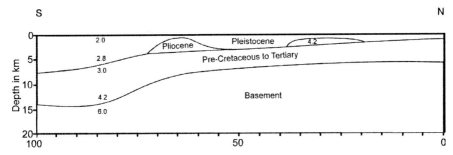

Fig. 4. Velocity-depth section along Dead Sea from north of the northern basin, to the southern basin. The 2.0 km/s velocity represents the Pleistocene fill of the basin. The 4.2 km/s velocity is associated with Pliocene evaporates. The 6.0 km/s velocity represents the top of the crystalline basement, while the overlying 3.0-3.8 km/s is associated with the Tertiary to pre-Cretaceous sediments (after Ginzburg, 1997)

the Pliocene and the Pleistocene affected river drainage patterns in the Negev and a massive rise in topography. Consequently large fresh water lakes developed in the southern Negev (such as Nahal Zihor, Avni, 1998; Ginat, 1996) and within the Arava valley, indicating a more humid environment (Avni, 1998). At the same time 200 m arching occurred in the Galilee (40-60 km wavelengths) dated at 1.8 Ma using basalt clasts constraint (Matmon et al., 1999).

Subsidence of the Dead Sea graben began in the early Miocene. Thick sequence of Pliocene and probably also Late Miocene evaporites, with Pleistocene sediments represent acceleration of subsidence rate (Fig. 5) (ten Brink and Ben-Avraham, 1989). Sedimentological evidence also supports a low topographic relief during the Pliocene, but a high relief during the Pleistocene (Sa'ar, 1985). A detailed study of subsidence rate in the southern Dead Sea basin was carried out by Larsen et al. (2002). It has been suggested (Ginzburg and Ben-Avraham, 1997) that during subsidence, normal faults between the northern and southern basins (Ben-Avraham, 1997) induced large vertical displacements, which in turn may have caused a greater deepening of the southern Dead Sea basin.

The tectonic processes that have modified the crustal structure along the Dead Sea fault are responsible for the creation of a morphological valley where a unique microclimate could develop. These ideal conditions would create a friendly environment, which would allow the migration of flora and fauna (and even man) from Africa northwards, within a wider corridor, often referred to as the Levantine Corridor (Bar-Yosef and Belfer-Cohen, 2001).

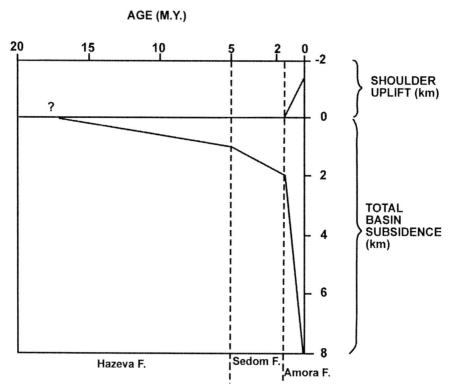

Fig. 5 Estimates for total subsidence of the Dead Sea basin with time based on stratigraphic interpretation of seismic lines. Approximation for the total up-lift is based on the current elevation of Late Cretaceous layers east of the Dead Sea basin (after ten Brink, 1989)

4. Ancient Hominids Out of Africa

The common approach in the study of ancient hominids and their culture, relates to the early archaeological evidence from eastern Africa, about 4-5 Ma, as the cradle of humankind (e.g. Bar-Yosef and Belfer-Cohen, 2001). The earliest evidence for human activity was found in Kenya, Ethiopia and Tanzania. Amongst these findings were remnants of stone tools related to the Acheulian (tool-making culture in which large nodules of flint were shaped to create hand axes), remnants of animals and even remains of hominids. These sites are dated as Plio-Pleistocene (>2 Ma). Hominid remains outside Africa delineate the routes of their spreading to the rest of the world (Fig. 6).

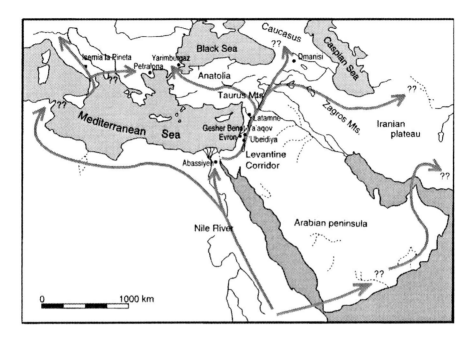

Fig. 6 Archeological evidence indicate at least three waves of early hominid migration out of Africa, but there were probably more. Map shows the sug-gested routes (after Bar-Yosef and Belfer-Cohen, 2001)

Human migration from Africa northwards is part of a wider phenomenon of migration of fauna and flora, which occurred along with the relief accentuation of the Dead Sea fault margins. A likely explanation for the relatively high number of tropical organisms, especially Ethiopian, along the Dead Sea fault is by migration, as today a vast desert separates them from their relatives in Africa and Asia (Tchernov, 1986). The DSF is a preferred migration route for billions of birds between Africa and Europe. It is one of the three major bird migration routes along with Gibraltar, and Sicily (Leshem, 1986), which are also known to be paths of hominid migration.

The morphology of the DSF created conditions in which fresh water bodies existed since the Pliocene. The lakes created friendly environments, richly varied in fauna and flora, for migrating hominids (Horowitz, 2001) along the fault. The Levantine Corridor, which extends from the Mediterranean coast on the west to the Jordanian plateau on the east, channeled hominids, technologies and materials from Africa to Asia and visa versa (Bar-Yosef, 1987). We argue that the DSF created a favorable zone within this corridor.

Archeological evidence of early hominid migration out of Africa show at least three pulses (Bar-Yosef and Belfer-Cohen, 2001). The earliest site in which human related flint artifacts were found is the Erk-el-Ahmar, about 10 km south of the Sea of Galilee (Fig. 1). Combined paleontologic and paleomagnetic dating of Erk-el-Ahmar yielded 1.7-2.0 Ma (Braun et al., 1991; Horowitz, 1979, 1989; Ron and Levi, 2001). The next pulses are observed at Ubediya, 1.4 Ma (Bar-Yosef and Goren-Inbar, 1993); and Gesher Benot-Ya'aqov, 0.78 Ma (Goren-Inbar et al., 2000).

In contrast to the sporadic findings in Erk-el-Ahmar, a series of sites were found in Ubediya (3 km southwest of the Sea of Galilee and 255 m below MSL), which represented numerous returns to the same location, close to a lake. According to Bar-Yosef and Belfer-Cohen (2001), the Ubediya site provides the best data set for the Levant. More than 60 archeological horizons were excavated in Ubediya, relating to different sedimentation sequences. The Acheulian artifact assemblage of Ubediya is very similar to the one in Upper Bed II of Olduvai Gorge (~1.4 Ma: Goren-Inbar, 2000). The bio-stratigraphic dating is based on the remains of over 100 species of mammals, birds and reptiles (Tchernov, 1999).

The mid-Acheulian site near Gesher Benot-Ya'aqov, on the banks of the Jordan River ~15 km north of the Sea of Galilee, dated to 0.78 Ma (paleomagnetic reversal and oxygen isotope stage 19), presents technological innovations, which appear for the first time outside of Africa (Goren-Inbar et al., 2000). The site is rich in flora and fauna remains. The Gesher Benot-Ya'aqov lithic assemblage, which is unique compared to contemporary sites in Eurasia, shows the development of complex human cognitive abilities in the tool industry.

5. Seismicity Along the Dead Sea Fault

"For thus saith the Lord of hosts; yet once, it is a little while, and I will shake the heavens, and the earth, and the sea, and the dry land" Haggai, 2:6.

The Dead Sea Fault is seismically active (Fig. 7). Man and earthquakes have coexisted along the DSF since early hominids migrated to this region some 2 Ma ago. Three sources provide information on past earthquakes: Instrumental data that have been amassed since the early 20th century, historical and archeological data that cover the last few millennia, and paleoseismic geological data that span tens of thousands of years. Paleoseismic studies along the DSF were initiated by Gerson and his colleagues in the

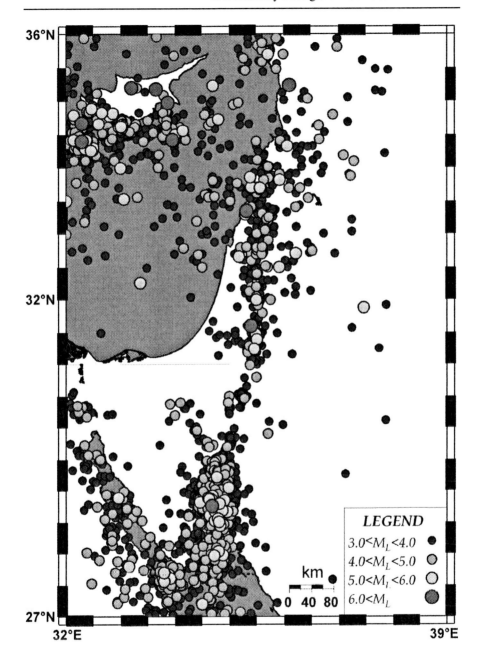

Fig. 7. Epicenters distribution map of earthquakes with ML>3 in the eastern Mediterranean between 1900-2000 (Geophysical Institute of Israel, 2002)

southern Arava (Gerson et al., 1993) and by Reches and Hoexter (1981) near Jericho. Subsequent studies have augmented our knowledge on the seismic activity of the DSF, although many questions still await further research.

The study of seismicity along the DSF benefits from several advantages. The tectonic framework is relatively simple with a single major plate boundary - the DSF. Arid climate entails excellent exposures. Sedimentary basins preserve sediments that potentially record tectonic events. In addition, the area has been inhabited throughout history. Written accounts on various phenomena, in particular earthquakes, are relatively abundant.

5.1 Paleoseismic Record

Broken and mixed lacustrine seasonal laminae in the Dead Sea basin ("mixed layers", Fig. 8) were interpreted as seismites - layers that exhibit earthquake-triggered deformation. Undisturbed laminated layers between these mixed layers represent interseismic intervals (Marco and Agnon, 1995). The mixed layers in the Late Pleistocene lacustrine Lisan Formation and its subsequent Dead Sea sediments comprise an almost continuous 70 ka paleoseismic record in the Dead Sea basin (Ken-Tor et al., 2001; Marco et al., 1996). Ken-Tor et al. (2001) show a remarkable agreement between mixed-layer ages and historical earthquakes in the last 2 millennia in the Dead Sea basin (Fig. 9), strengthening the interpretation of the mixed layer as seismites. Enzel et al. (2000) recovered evidence of combined faulting and shaking effects in the Darga alluvial fan, northern Dead Sea. Paleoseismic on-fault studies revealed slip histories on marginal normal faults of the southern Arava (Amit et al., 2002), the Dead Sea (Marco and Agnon, 2002), and the Hula basin (Zilberman et al., 2000). Strike-slip movements have been measured and dated on the Jordan Gorge Fault, north of the Sea of Galilee, where 2.2 m and 0.5 m sinistral slip occurred in the earthquakes of 1202 and 1759 respectively. A 15 m displacement of a 5 ka stream channel gives a minimum average displacement rate of 3 mm/a (Marco and Agnon, 2003). Klinger et al. (2000a) and Niemi et al. (2001) dated offset channel deposits in alluvial fans in the northern Arava. They conclude an average Late Quaternary slip rate of about 4-5 mm/a (Table 1). For comparison, it is an order of magnitude smaller than across the San Andreas system.

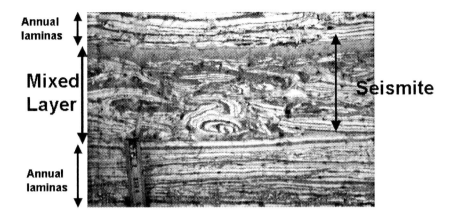

Fig. 8. Broken and mixed lacustrine seasonal laminae in the Dead Sea basin ("mixed layers"), which are interpreted as seismites - layers that exhibit earthquake-triggered deformation. Undisturbed laminated layers between these mixed layers represent interseismic intervals

Table 1. Slip rate estimates of the Dead Sea Fault

Period	Rate [mm/a]	Data	Reference
Post Miocene	6 [0.283°/ma]	Plate kinematics	Joffe and Garfunkel, 1987
Plio-Pleistocene	7-10	Geological	Garfunkel et al., 1981
Plio-Pleistocene	20	Geological	Steinitz and Bartov, 1986
Plio-Pleistocene	5.4-6.1	Geological	Heimann, 1990
Plio-Pleistocene	3-7	Drainage systems, Arava Fault	Ginat et al., 1998
Pleistocene	2-6, prefer 4	Alluvial fans, N. Arava	Klinger et al., 2000a; Klinger et al., 2000b
Pleistocene	4.7±1.3	Alluvial fans, Arava	Niemi et al., 2001
Late Pleistocene	6.4±0.4	Seismicity	El-Isa and Mustafa, 1986
Late Pleistocene-Recent	10	Geological	Freund et al., 1968
12 ka	4.5-4.8	Geological, alluvial fans, N. Arava	Zhang et al., 1999
Holocene	9	Geological	Reches et al., 1987
Holocene	0.7	Geological	Gardosh et al., 1990
Last 5000 a	3	Stream channel, Jordan Gorge	Marco et al., 2000
Last 4500 a	2.2	Seismicity	Ben-Menahem, 1981
Last 1000 a	0.8-1.7	Historical	Garfunkel et al., 1981
Last 800 a	<2.5	Archaeological	Marco et al., 1997

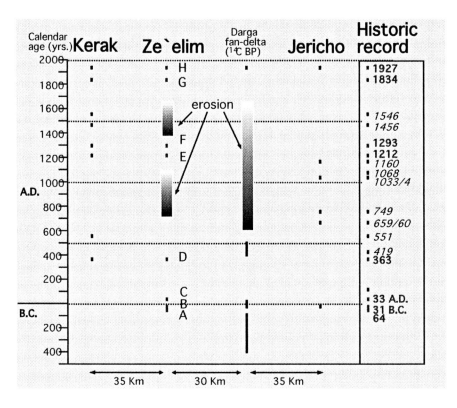

Fig. 9 Correlation between mixed-layer ages and historical earthquakes in the last 2 millennia in the DS basin at four locations (after Ken-Tor et al., 2001). C14 – ages of mixed layers A-H in Ze'elim are correlated with the historic record of earthquakes in the area (right column dates in bold were measured in Ze'elim). The earthquakes reported either from Karak (35 km to the southeast) and/or from Jericho (some 60 km north of Ze'elim)

5.2 Historical and Archeological Earthquake Record

As described above, the Dead Sea Fault is part of the hominid migration route out of Africa. Some of the settlements have been affected by earthquakes and the observed damage together with historical accounts, provide a unique record of past seismic activity in the region (e.g. Amiran et al., 1994).

Two forms of written texts of natural catastrophes can be found: biblical stories where uncertainty is high, such as the destruction of Jericho (Joshua

Jerash 749 AD   Sussita 749 AD

Kal'at Nemrod 1759   Jericho 1927

Fig. 10. Ruins of Jerash, Kalaat Nemrod, Jericho and Sussita, which were damaged by earthquakes during the last 2000 years (see Figure 2 for locations).

6) and contemporary detailed accounts, such as Josephus Flavius' vivid description of the 31BC earthquake. Archeology can usually corroborate history although interpreting damage to ancient structures is not trivial. Damage related to earthquake shaking is recognized in numerous ruins, including Jerash, Jericoh, Sussita, and Kalaat Nemrod (Fig. 10). Finding evidence for historical earthquake ruptures is extremely rare. Recently, the first such evidence from the DSF has been recovered from the Crusader fortress of Vadum Iacob (now called Ateret near Gesher Benot Ya'aqov – Fig. 2) (Ellenblum et al., 1998; Marco et al., 1997b). The fortress, which was built on the active trace of the fault, was torn apart twice, first by the earthquake of 1202 and again by the earthquake of 1759. Both events are in agreement with the paleoseismic observations mentioned above. A water reservoir at the Roman-Early Byzantine site Kasr-e-Tilah in the northern Arava was also offset by the fault (Klinger et al., 2000b).

5.3 Instrumental Record

The seismological division of the Geophysical Institute of Israel operates some 100 monitoring systems of the modern Israel Seismic Network. The accrued data reveal several important characteristics.

most active area has been the Gulf of Elat, where thousands of small earthquakes cluster during periods of several months to a few years in different regions of the gulf. The activity culminated in the 22 Nov. 1995 Mw7.1 earthquake (Baer et al., 1999; Klinger et al., 1999).

Focal plane solutions of other events confirm the primary sinistral motion predicted by local geology and plate tectonic considerations. Local complications in the form of stepovers are manifested by normal faulting. Reverse fault solutions are rare (Salamon et al., 1996; van-Eck and Hofstetter, 1989, 1990). The earthquakes largely obey the Gutenberg-Richter magnitude-frequency distribution with typical b values of 0.85-0.9 (Shapira and Shamir, 1994).

5.4 Patterns of Seismicity

By combining several disciplines including history, archaeology, and seismology, several spatio-temporal patterns begin to emerge from the data. Clustering, periodicity, and triggering have been reported in several studies. In accordance with theoretical and experimental results (e.g., Lyakhovsky, 2001), a temporal pattern in a century time window resembles that of the 10 ka window. The longest continuous off-fault record (68-18 ka) from the lacustrine Lisan Formation shows that strong (M>6) earthquakes cluster during periods of ~10 ka, with more quiet periods between the clusters. During the long cluster periods earthquakes appear in secondary clusters (Marco et al., 1996). A pattern of clustering is also evident in the record of the last two millennia in the Dead Sea (Ken-Tor et al., 2001) and in the M>4 record of the 20th century (Marco and Agnon, 2001)

Bearing in mind the large uncertainties associated with the interpretation of historical accounts, it seems that a unique quasi-periodic recurrence of four large earthquakes occurred in the Jordan Valley between the Dead Sea and the Sea of Galilee from 31 BC through AD 363, 749, and 1034. This sequence was followed by the two smaller earthquakes of 1546 and 1927 with a similar recurrence interval (Marco and Agnon, 2001). The quasi-periodic pattern appears to last only for a short period on the order of a few centuries, and is not observed elsewhere along the DSF. An alarming note is the absence of strong (M>6.5) earthquakes in the last eight centuries between Lebanon and the Dead Sea, since most of the DSF sustained rupture

during the 1034 earthquake in the Jordan Valley and the 1202 earthquake
north of the Sea of Galilee.

5.5 Example: Jericho

"Jericho is the latch of the Land of Israel. If Jericho was taken the whole
country would instantly be conquered" Rabbi Samuel Bar-Nahmani (AD
426-500), Midrash Rabba.

Settlements along the Dead Sea Fault have been repeatedly affected by
earthquakes. An example can be found in the ancient city of Jericho. Tectonics has played an important role in the cultural record of Jericho. On
one hand, its unique position close to an active strand of the Dead Sea
transform, gives rise to a source of fresh water – the Spring of Elisha,
which probably ascends through channels along fractured rocks east of the
main fault (Neev and Emery, 1995). The abundance of fresh water in such
an arid area, has led Jericho to play an important role in history – that of
the oldest continuously settled city in the world. The first human settlement was by Mesolithic people of the Natufian age, 12,500-10,000 years
ago who were probably associated with the first Mesolithic group further
along the transform – at the Mount Carmel caves (Kenyon, 1960). Jericho
is one of the first places where evidence of man as a member of a settled
community and a food producer rather than a food gatherer, exists (Pre-
Pottery Neolithic A - Kenyon, 1960). In addition, the settlement at Jericho
is around two thousand years older than the earliest known villages elsewhere. The spring has also allowed inhabitants of the city to resist siege,
by providing a constant supply of vital water, thereby limiting the need for
abandonment due to destruction or capture. On the other hand, what man
could not accomplish, nature could.

The proximity to an active fault meant that Jericho has been faced with
multiple destructions due to earthquakes. Damage to the city's outer walls,
is well documented (Kenyon, 1960) and the city itself has been destroyed
and rebuilt at least 17 times. This view is supported by paleoseismological
data (Reches and Hoexter, 1981). Radiocarbon dating has shown that de-
fensive walls existed around Jericho as far back as 7000 BC (Kenyon,
1960). The most famous account of destruction by earthquake is of course,
the biblical tale.

Ancient Jericho was the meeting place of several important trade routes
and thus a strategically important place to control. The biblical importance
of the city cannot be overlooked. Jericho, the first town encountered in the
Promised Land after the exodus from Egypt, was the gateway to Israel.
The narrative reports that the waters of the Jordan River stopped flowing

and allowed the Israelites to pass (Joshua 3:13-16). Earthquakes are known to have caused landslides, which dam the river and interrupt its flow for several hours or even days. Reports of such events exist from 1160, 1267 (16 hours), 1546, 1834, 1906 and 1927 (21 hours) (Amiran et al., 1994; Ben-Menahem, 1981). The destruction of the walls of the city and the damming of the river, as described in Joshua 6:1-16, is generally agreed by most archeologists to be the result of an earthquake, possibly on the Jericho Fault (Neev and Emery, 1995).

In 31 BC, during the 7th year of King Herod's reign, a strong earthquake hit the region. The historian Josephus Flavius records: "...*and there was an earthquake in Judea, such as had not occurred before, which killed many cattle throughout the country. And about thirty thousand persons also perished in the ruins of their houses, but the army, which lived in the open, was not at all harmed by this calamity*". The Arabs, believing that Judea was in ruins due to the earthquake, intended to invade it. However, King Herod managed to lead his army across the Jordan River and defeated the Arabs. Consequently, as a result of the earthquake his kingdom was enlarged.

6. Discussion

Active tectonic processes along the DSF have fashioned an environment that has influenced the course of human history as the critical bridge between the continents of Africa, Europe and Asia, creating a corridor of friendlier environments through which hominids migrated. The region has played an important role in Near Eastern prehistory and archeology. The historical and archeological associations of this area are extensive. Here is where the oldest sites of hominids outside Africa, Erk-el-Ahmar and Ubediya are located, as well as Jericho, the oldest city in the world.

The uplift that created the unique physiography of the DSF is controlled by the crustal structure and the tectonic processes in this region. The crust was formed during the processes that shaped the Arabian-Nubian shield in the Precambrian and later modified by the formation of the DSF plate boundary. The precise location of the DSF was suggested to be on a rejuvenation of an old Precambrian weakness zone (Girdler, 1991). Garfunkel and Ben-Avraham (2001) showed that the mechanical properties of the Precambrian basement may have affected the various styles of deformation along the DSF margins, but the general trace of the fault is totally different from the old structures. It is therefore highly unlikely that the DSF is a rejuvenated Precambrian structure.

Earthquakes that are associated with the Dead Sea Fault plate boundary and also reflect the crustal structure because they occur where the crust fails due to regional stresses. Lyakhovsky et al. (1997) relate the failure to growth patterns of distributed damage in the lithosphere. Once a fault zone is created, the occurrence of earthquakes is influenced by the geometry of the fault zone, which evolves in time from complex to simple geometry (Stirling et al., 1995) and by the velocity of the plate movements. Hence the occurrence of earthquakes is governed by the detailed combination of crustal structure and regional motions of the tectonic plates. The theoretical work by Lyakhovsky et al. (2001) suggests that the relatively slow plate velocity along the Dead Sea Fault is compatible with long-term clustering.

The immediate effects of damaging earthquakes on society are known and familiar to modern people. Unfortunately earthquake-inflicted human tragedies and economic losses are huge. It is less clear how earthquakes affected society in the past and whether they significantly changed the course of history. For example the 31 BC earthquake in the Dead Sea region caused confusion and fear, which paved the way for the expansion of Herod's kingdom. Episodic time-space clustering of earthquakes such as during the eastern Mediterranean 'seismic crisis' at the end of the Bronze Age (around 1200 BC) and in the 4th century AD (Nur, 2000). The seismicity of the north Anatolian fault during the 20th century is another example of a cluster of strong earthquakes. Society benefits as well as is harmed by the same geological processes. Studying the ways in which earthquake affected societies may help us cope with such catastrophies that will certainly occur again.

Acknowledgements

The authors wish to thank Yossi Leshem, Aharon Horowitz, and Ravid Ekshtain for their help and comments.

References

Amiran DHK, Arieh E, Turcotte T (1994) Earthquakes in Israel and adjacent areas: Macroseismic observations since 100 B.C.E. Israel Exploration Journal 44:260-305

Amit R, Zilberman E, Enzel Y, Porat N (2002) Paleoseismic evidence for time dependency of seismic response on a fault system in the southern Arava valley, Dead Sea rift Israel. Geological Society of America Bulletin 114:192-206

Avni Y (1998) Paleogeography and tectonics of the central Negev and the Dead Sea rift western margin during the Late Neogene and Quaternary. Geological Survey of Israel, GSI/24/98

Baer G, Sandwell D, Williams S, Bock Y, Shamir G (1999) Coseismic deformation associated with the November 1995, Mw = 7.1 Nuweiba earthquake, Gulf of Elat (Aqaba), detected by synthetic aperture radar interferometry. Journal of Geophysical Research 104:25,221-25,232

Bar-Yosef O (1987) Pleistocene connexions between Africa and southeast Asia: an archaeological perspective. The African Archaeological Review 5:29-38

Bar-Yosef O, Goren-Inbar N (1993) The lithic assemblages of 'Ubadiya'. Qedem - Monographs of the Institute of Archaeology 34 Institute of Archaeology, Hebrew University, Jerusalem

Bar-Yosef O, Belfer-Cohen A (2001) From Africa to Eurasia - early dispersals. Quaternary International 75:19-28

Ben-Avraham Z (1997) Geophysical framework of the Dead Sea: structure and tectonics. In: Niemi TM, Ben-Avraham Z, Gat JR (eds) The Dead Sea, The Lake and Its Setting. Oxford University Press, New York pp. 22-35

Ben Avraham Z (2001) The Dead Sea - a unique global site. European Review 9:437-444

Ben-Menahem A (1981) Variation of slip and creep along the Levant Rift over the past 4500 years. Tectonophysics 80:183-197

Braun D, Ron H, Marco S (1991) Magnetostratigraphy of the homonid tool-bearing Erk el Ahmar formation in the northern Dead Sea Rift. Israel Journal of Earth Sciences 40: 191-197

DESERT Group (2000) Multinational geoscientific research effort kicks off in the Middle East. Eos Transactions, AGU 81: 609, 616-617

El-Isa ZH, Mustafa H (1986) Earthquake deformations in the Lisan deposits and seismotectonic implications. Geophysical Journal Royal Astronomical Society 86:413-424

Ellenblum R, Marco S, Agnon A, Rockwell T, Boas A (1998) Crusader castle torn apart by earthquake at dawn, 20 May 1202. Geology 26:303-306

Enzel Y, Kadan G, Eyal Y (2000) Holocene earthquakes inferred from a fan-delta sequence in the Dead Sea graben. Quaternary Research 53:34-48

Freund R, Zak I, Garfunkel Z (1968) Age and rate of the sinistral movement along the Dead Sea Rift. Nature 220:253-255

Gardosh M, Reches Z, Garfunkel Z (1990) Holocene tectonic deformation along the western margins of the Dead Sea. Tectonophysics 180:123-137

Garfunkel Z (1981) Internal structure of the Dead Sea leaky transform (rift) in relation to plate kinematics. Tectonophysics 80:81-108

Garfunkel Z, Ben-Avraham Z (2001) Basins along the Dead Sea transform: Peri-Tethys Memoir 6: Peri-Tethyan rift/wrench basins and passive margins. Memoires Du Museum National D'Histoire Naturelle 186:607-627

Garfunkel Z, Zak I, Freund R (1981) Active faulting in the Dead Sea rift. Tectonophysics 80:1-26

Geophysical Institute of Israel (2002) Epicenter distribution map of earthquakes. http://www.gii.co.il

Gerson, R, Grossman S, Amit R, Greenbaum N (1993) Indicators of faulting events and periods of quiescence in desert alluvial fans. Earth Surface processes and landforms 18:181-202

Ginat H, Enzel Y, Avni Y (1998) Translocation of Plio-Pleistocene drainage system along the Dead Sea Transform, south Israel. Tectonophysics 284:151-160

Ginzburg A, Ben-Avraham Z (1997) A seismic refraction study of the north basin of the Dead Sea, Israel. Geophysical Research Letters 24:2063-2066

Ginzburg A, Makris J, Fuchs K, Prodehl C, Kaminski W, Amitai U (1979) A Seismic study of the crust and upper mantle of the Jordan-Dead Sea rift and their transition toward the Mediterranean Sea. Journal of Geophysical Research 84:1569-1582

Ginzburg A, Makris J, Fuchs K, Prodehl C (1981) The structure of the crust and upper mantle in the Dead Sea rift. Tectonophysics 80(1-4):109-119

Girdler RW (1991) The Afro-Arabian rift system - An overview. Tectonophysics 197:139-153

Goren-Inbar N, Feibel CS, Verosub KL, Melamed Y, Kislev ME, Tchernov E, Saragusti I (2000) Pleistocene milestones on the Out-of-Africa Corridor at Gesher Benot Ya'aqov, Israel. Science 289(5481):944-947

Heimann A (1990) The development of the Dead Sea rift and its margins in the northern Israel during the Pliocene and the Pleistocene. Golan Research Institute and Geological Survey of Israel GSI/28/90

Horowitz A (1979) The Quaternary of Israel. Academic Press, New-York

Horowitz A (1989) Continuous pollen diagrams for the last 3.5 my from Israel: vegitation, climate and correlation with oxygen isotope record. Paleogeogr Paleoclimatol Paleoecol 72: 63-78

Horowitz A (2001) The Jordan rift valley. A. A. Balkema Publishers, Lisse

Joffe S, Garfunkel Z (1987) Plate kinematics of the circum Red Sea- a re-evaluation. Tectonophysics 141:5-22

Ken-Tor R, Agnon A, Enzel Y, Marco S, Negendank JFW, Stein M (2001) High-resolution geological record of historic earthquakes in the Dead Sea basin. Journal of Geophysical Research 106:2221-2234

Kenyon K (1960) Archaeology in the Holy Land. Fredrick A. Praeger, Inc, New York

Klinger Y, Rivera L, Haessler H, Maurin JC (1999) Active faulting in the Gulf of Aqaba: New knowledge from the Mw7.3 earthquake of 22 November 1995. Bulletin of the Seismological Society of America 89:1025-1036

Klinger Y, Avouac JP, Abou-Karaki N, Dorbath L, Bourles D, Reyss JL (2000a) Slip rate on the Dead Sea transform fault in northern Araba Valley (Jordan). Geophysical Journal International 142:755-768

Klinger Y, Avouac JP, Dorbath L, Abou-Karaki N, Tisnerat N (2000b) Seismic behavior of the Dead Sea Fault along Araba Valley, Jordan. Geophysical Journal International 142:769-782

Larsen B, Ben-Avraham Z, Shulman H (2002) Fault and salt tectonics in the southern Dead Sea basin. Tectonophysics 346:71-90

Leshem Y (1986) The Dead Sea Rift as an intercontinental axis of foaring birds. Rotem: The evolution of the Dead Sea Rift: Ecology, Botany, Prehistory pp. 118-134

Lyakhovsky V, Ben-Zion Y, Agnon A (1997) Distributed damage, faulting, and friction. Journal of Geophysical Research 102:27,635-27,649

Lyakhovsky V, Ben-Zion Y, Agnon A (2001) Fault evolution and seismicity patterns in a rheologically layered halfspace. Journal of Geophysical Research 106:4103-4120

Marco S, Agnon A (1995) Prehistoric earthquake deformations near Masada, Dead Sea graben. Geology 23:695-698

Marco S, Agnon A (2003) Repeated earthquake faulting revealed by high-resolution stratigraphy. Tectonophysics (in press)

Marco S, Stein M, Agnon A, Ron, H (1996) Long term earthquake clustering: a 50,000 year paleoseismic record in the Dead Sea Graben. Journal of Geophysical Research 101:6179-6192

Marco S, Agnon A, Ellenblum R, Eidelman A, Basson U, Boas A (1997a) 817-year-old walls offset sinistrally 2.1 m by the Dead Sea Transform, Israel. Journal of Geodynamics 24:11-20

Marco S, Agnon A, Stein M, Bruner I, Landa E, Basson U, Ron H (1997b) A 70 kyr paleoseismic record in the Dead Sea graben recovered by combined geophysical and geological studies. The 13th GIF meeting on the Dead Sea Rift as a unique global site, Dead Sea, Israel

Matmon A, Enzel Y, Zilberman E, Heimann A (1999) Late Pliocene to Pleistocene reversal of drainage systems in northern Israel: tectonic implications. Geomorphology 28:43-59

Neev D, Emery KO (1995) The destruction of Sodom, Gomorrah, and Jericho: Geological, climatological and archaeological background, Oxford University Press, Oxford

Niemi TM, Zhang H, Atallah M, Harrison BJ (2001) Late Pleistocene and Holocene slip rate of the Northern Wadi Araba fault, Dead Sea Transform, Jordan. Journal of Seismology 5:449-474

Nur A, Cline EH (2000) Poseidon's horses: Plate tectonics and earthquake storms in the Late Bronze Age Aegean and Eastern Mediterranean. Journal of Archaeological Science 27:43-63

Reches Z, Hoexter DF (1981) Holocene seismic and tectonic activity in the Dead Sea area Tectonophysics 80:235-254

Reches Z, Erez J, Garfunkel Z (1987) Sedimentary and tectonic features in the northwestern Gulf of Elat, Israel. Tectonophysics 141:169-180

Ron H, Levi S (2001) When did homonids first leave Africa?: New high-resolution magnetostratigraphy from the Erk el Ahmar formation, Israel. Geology: 29(10):887-890

Sa'ar H (1985) Origin and sedimentation of sandstones in graben fill formations of the Dead Sea rift valley. Report MM/3/86 Geological Survey of Israel, Jerusalem

Salamon A, Hofstetter A, Garfunkel Z, Ron H (1996) Seismicity of the eastern Mediterranean region: Perspective from the Sinai subplate. Tectonophysics 263:293-305

Shapira A, Shamir G (1994) Seismicity parameters of seismogenic zones in and around Israel. Institute for Petroleum Research and Geophysics Report Z1/567/79

Steinitz G, Bartov Y (1986) The 1985 time table for the tectonic events along the Dead Sea transform. Terra Cognita 6:160

Stirling MW, Wesnousky SW, Shimazaki K (1995) Fault trace complexity, cumulative slip, and the shape of the magnitude-frequency distribution for strike-slip faults: A global survey. Geophysical Journal International 124:833-868

Tchernov E (1986) The history of the fauna in the Dead Sea Rift and its biogeographical meaning. Rotem: The evolution of the Dead Sea Rift: Ecology, Botany, Prehistory p.51-75

Tchernov E (1995) The earliest hominids in the southern Levant. In: Proceedings of the International Conference of Human Palaeontology, Orce, Spain, p. 389-406

ten Brink US, Ben-Avraham Z (1989) The anatomy of a pull-apart basin: seismic reflection observations of the Dead Sea basin. Tectonics 8(2):333-250

ten Brink US, Schoenburg N, Kovach RL, Ben-Avraham Z (1990) Uplift and possible Moho offset across the Dead Sea transform. Tectonophysics 180:77-92

van-Eck T, Hofstetter A (1989) Microearthquake activity in the Dead Sea region. Geophysical Journal International 99:605-620

van-Eck T, Hofstetter A (1990) Fault geometry and spatial clustering of microearthquakes along the Dead Sea-Jordan rift fault zone. Tectonophysics 180:15-27

Wdowinski S, Zilberman E (1997) Systematic analyses of the large scale topography and structure across the Dead Sea Rift. Tectonics 16:409-424

Zhang H, Niemi TM, Atallah M, Harrison BJ (1999) Slip rate of the northern Wadi Araba Fault, Dead Sea Transform, Jordan during the past 12,000 years. In: Geological Society of America annual meeting, Boulder, CO., p. 114

Zilberman E, Amit R, Heimann A, Porat N (2000) Changes in Holocene paleo-seismic activity in the Hula pull-apart basin, Dead Sea Rift, northern Israel. Tectonophysics: 321:237-252

Velocity Field of the Aegean-Anatolian Region from 3D Finite Element Models

Oliver Heidbach

Heidelberg Academy of Sciences and Humanities, "World Stress Map Project"
Geophysical Institute, Karlsruhe University

Abstract

In order to investigate the geodynamic processes and rheological parameters, which are principally responsible for the observed surface deformation, a three-dimensional finite element model of the Eastern Mediterranean was built up. The northward movement of the Arabian and the African plate relative to the Eurasian plate has been assumed to control the tectonic evolution of the Aegean-Anatolian block since the Miocene. The indentation of Arabia forces the Anatolian region to escape westwards along the North and the East Anatolian fault, whilst the Aegean region is extended by southward retreat of the Hellenic arc subduction zone. Major faults were modelled as contact surfaces with Coulomb friction. Boundary conditions imposed were the slab pull beneath the Hellenic arc and the northward movement of the African and the Arabian plate using a rigid plate model. An elasto-visco-plastic rheology was applied for the modeling time of 0.5 Ma. By varying the properties a set of parameters was determined which gives a minimal mean deviation between the modeled velocity field and GPS data from 45 observation sites. This comparison is considered justified, since even though the geodetic observations are representing a short time period, the geodetically derived displacement rates coincide with long-term rates derived from seismicity and fault-slip data. The best-fitting model was attained with the following parameters: (a) friction coefficient between 0.2 and 0.45 along the subduction zones and the main faults, (b) slab pull stresses of 150 MPa, (c) cohesive strength of 460 MPa in Anatolia (compressional regime) and (d) cohesive strength of 40 MPa in the Aegean (extensional regime). In this case the mean deviation was 21% in magnitude and $\pm 10°$ in azimuth. The strain rate field of the model has a minimum in the Anatolian region ($< 2 \cdot 10^{-15}$ s^{-1}) and a maximum in the Aegean region (5-10$\cdot 10^{-15}$ s^{-1}). These values are comparable with the strain rates derived from seismicity. The velocity field of the best-fitting model shows almost rigid westward movement of Anatolia and SSW oriented extension in the Aegean region.

1. Introduction

Due to the ongoing convergence between the African, Arabian and Eurasian plates a complex tectonic situation has developed in the Eastern Mediterranean. Subduction zones, strike-slip faults and continental collision zones alternate one another. They form broad deformation belts along the borders of the plates (Kahle and Mueller 1998; Jolivet and Faccenna 2000). As a first order approximation the velocity field can be described through the movement of rigid plates and blocks (DeMets et al. 1994; McKenzie 1972). The movement of the Anatolian-Aegean block is determined by the major plates around this block. It is bordered by the North and the East Anatolian fault, the subduction zones along the Hellenic and the continental collision at the Cypriotic arc (Fig. 1). However, the description of the velocity field with a rigid body model is not capable to explain the internal deformation and the mechanisms which drive the block movement and control the deformation. A continuous deformation model has to be defined whose partial differential equations can often only be computed by numerical algorithms.

The existing 2-D finite element deformation models of the Eastern Mediterranean use an elastic rheology (e.g. Meijer and Wortel 1997; Lundgren et al. 1998) or power-law rheology (e.g. Cianetti et al. 1997). The presented models extend to 3-D in order to avoid a horizontal parameterization of the slab pull stresses, to take into account the internal forces due to lateral density variations and to investigate the effect of rheological layering on the surface deformation. The velocity field of each model is compared with Global Positioning System (GPS) observations at 45 sites on the Aegean-Anatolian block published by McClusky et al. (2000).

2. Tectonic Development of the Eastern Mediterranean

In the Eastern Mediterranean the African plate moves with 0.9 cm/a in northern direction relative to a fixed Eurasian plate (DeMets et al. 1994). Due to the opening of the Gulf of Aden the Arabian plate separated from the African plate along the Dead Sea fault system during middle Miocene and started an independent northward movement with 2.5 cm/a (DeMets et al. 1994; Westaway 1994; Robertson and Grasso 1995). The westward escape of Anatolian started along the North and the East Anatolian fault at the end of Miocene, when continental collision began at the northern edge of the Arabian plate.

During late Miocene and Pliocene the length of the subduction zones in the Mediterranean have shortened. Today the oceanic crust is already completely consumed by subduction west of the Malta Escarpment and probably east of Crete while a substantial part south of the Hellenic arc and in the Ionian sea is still present (Robertson 1998; Reuther et al. 1993). After continental collision started, the still ongoing gravitational forces cause increasing tensional stresses inside the subducted slabs. When the tensional strength is reached the subducted slabs start to detach. This process is proposed by several investigations using seismic tomography data (e.g. Wortel and Spakman 1992; de Jonge et al. 1994; Spakman et al. 1993). Slab detachment could have happened west of the Malta escarpment, under the Apennines north of the Gulf of Tarent, under the Dinarides north of the Kephalonia fault and under the Cypriotic arc. It is still under debate whether subduction beneath at the Cypriotic arc continues and at which location it occurs (Ben-Avraham et al. 1988; Ben-Avraham and Ginzburg 1990).

A relict of approximately 90 Ma old oceanic crust is preserved in the Ionian sea between the two subduction zones of the Calabrian and Hellenic arc (Mueller and Kahle 1993). The Calabrian arc subduction zone is 300-400 km long and sharply bended between the Malta escarpment in the south-west and the Gulf of Tarent in the north(Fig. 1). The Hellenic arc subduction zone starts south of the dextral Kephalonia fault and terminates near the sinistral Strabo-Plini strike-slip fault system. Probably since the beginning of Miocene both subduction zones are retreating. This migration lead to back-arc extension in the Tyrrhenian Sea and the Aegean Sea.

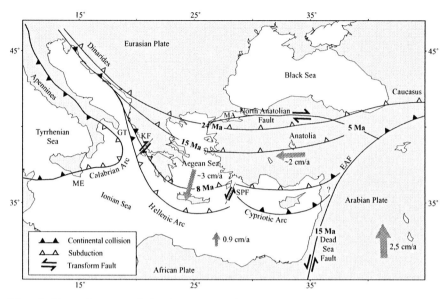

Fig. 1. Tectonic map of the Eastern Mediterranean. The gray arrows give the movement directions of the Arabian and the African plate and of the Anatolian-Aegean block relative to the fixed Eurasian plate. (EAF=East Anatolian Fault, GT=Gulf of Tarent, KF=Kephalonia Fault, MA=Marmara Sea, ME=Malta Escarpment, SPF=Strabo-Plini Fault). Age dates in Ma display the southward migration of the Hellenic arc subduction zone and the onset time of the major strike-slip faults (Ziegler 1988; Meulenkamp 1988 and Dewey et al. 1989).

3. Geophysical, Geological and Geodetic Observations

The westward movement of Anatolia and the south-southwest movement of the Aegean Sea is nowadays observed with high accuracy by means of GPS (Fig. 2). These observations provide information about the movement of discrete points from campaign and permanent measurement. Although the observation time covers less then a decade the observed velocity field differs only slightly from other observations method as for instance palaeomagnetic studies or fault-slip analysis (Westaway 1994).

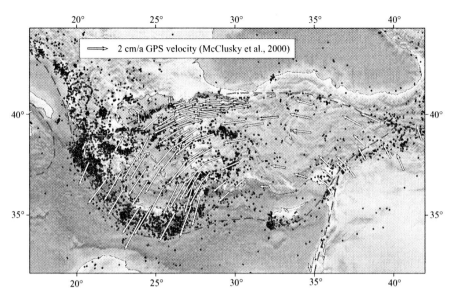

Fig. 2. Seismicity from 1900-2000 with magnitude >4.0 and depth <40 km (data from the BGR Hannover http:\\www.bgr.de) and results from GPS observations in the Eastern Mediterranean (McClusky et al. 2000). Velocities are presented relative to the fixed Eurasian plate reference system.

In the Aegean region high seismicity can be observed inside the broad deformation belts along its borders. For the Anatolian region only the western part shows high seismicity whilst in the central and eastern part it is relatively low. From the seismological strain rate (Jackson and McKenzie 1988; Jackson et al. 1992) and from geological indicators (Mercier et al. 1987; Meulenkamp et al. 1988) it is known that Anatolia moves mainly rigidly with only small internal deformation. The Aegean region is deformed by extension which is also expressed through a thinned crust. The crustal thickness in the western part of Anatolia is 30-35 km compared to the central Aegean Sea where it decreases to 20-25 km (Geiss 1987; Makris and Stobbe 1984). This crustal thickness variation of about 10 km induces a horizontal pressure gradient. Earthquake focal mechanisms in the northern part of the Aegean region reflect extension and strike-slip. Along the North and the East Anatolian fault mainly strike-slip occurs. Near the Hellenic and the Cypriotic subduction zone thrusting dominates (Ben-Avraham et al. 1995; Taymaz et al. 1990). Further information on the tectonic stress field are compiled in the World Stress Map database (Reinecker et al. 2003, Sperner et al. 2003).

4. Model Geometry

The model geometry has to allow for relative movements between the main blocks. These movements take place along the Hellenic arc, the Cypriotic arc, the Dead Sea fault, the North and the East Anatolian fault system, and the border between the Arabian and the European plate. Small-scale tectonic features had to be neglected since they can not be resolved in such large-scale models. Variations in Moho depth and lithospheric thickness were allowed for. Data were taken from Geiss (1987) and Heidbach (2000). As well topography is included. Relative movements between the plates and blocks were modelled with contact surfaces where Coulomb friction is defined.

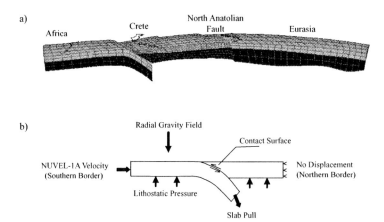

Abb. 3. a) Block diagram of the finite element model. The gray color indicates the crust and the black color the lithospheric part of the upper mantle. **b)** View of a cross section with the boundary conditions. NUVEL-1A is the global plate from DeMets et al (1994).

5. Model Assumptions and Boundary Conditions

The three models presented in this paper are based on the following assumptions:

(1) Accelerations are neglected, since gravity effects are much higher than accelerations of the tectonic processes. (2) Incompressibility is assumed (Boussinesq approximation). In case of steady-state creep, which is reached in the model time of 0.5 Ma, the change of the rock volume is neg-

ligible small. The effect of the elastic part of the total deformation (elastic+plastic+creep) is small. Therefore the volume change due to elastic compression in the model is negligible. (3) Constant thermal properties. (4) Spherical earth, i.e. no membrane stresses due to north or south movement. (5) No phase transitions or other chemical processes during subduction.

Boundary conditions applied in the models are: (1) Spherical gravity field. (2) Lithostatic pressure at the bottom of the model. No shear forces are applied at the bottom of the model following the idea that the viscosity contrast between the lithosphere and the asthenosphere is too high to produce a reasonable coupling.

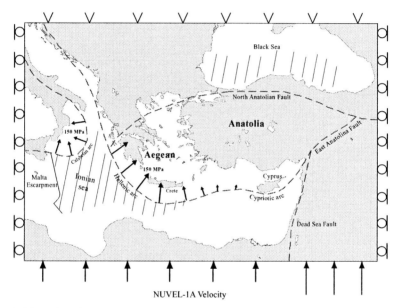

Fig. 4. Map sketch of the model boundary conditions. Hatched areas mark the location of 90 Ma old oceanic crust from the Neotethys and the Black Sea. The northern border of the model is fixed, sides are allowed to roll in North-South direction and at the southern border NUVEL-1A plate velocities from DeMets et al. (1994) are applied. Dashed lines indicate the location of the implemented contact surfaces with Coulomb friction.

(3) Kinematic constraints from global rigid plate model NUVEL-1A (DeMets et al. 1994) at the models southern border. (4) Fixed northern border of the model and free slip along the models eastern and western borders in N-S direction, i.e. movement in E-W direction at the models sides is suppressed. (6) Slab-pull stresses in the direction of the subducting slabs. At the Cypriotic arc and north of the Kephalonia fault no slab pull

stresses are assumed since slab detachment has already occurred and continental collision processes are taking place at present. The indentation of the Arabian plate and the ongoing subduction beneath the Hellenic arc are assumed to be the main driving forces for the observed surface deformation and the movement of the Anatolian-Aegean block (Fig. 4).

6. Model Rheology and Material Properties

The model behaves as an extended Bingham body with Maxwell rheology (Fig. 5). The rheology of the upper crust is described with an elastic-plastic body. Its yield strength σ_y is computed using the Mohr-Coulomb criterion

$$\sigma_y = \mu \cdot \sigma_n + C \qquad (1)$$

where C is the cohesive strength, σ_n is the normal stress and μ the friction coefficient. The rheology of the lower crust and the lithospheric mantle is described with a visco-elastic body.

In order to simulate a plastic behavior, stresses in the upper crust which are above the yield strength are not relaxed with an instantaneous ideal plastic process, but with a retardation according to a relatively low viscosity of 10^{18} Pa s. The lower crust and the upper mantle have no yield strength, i.e. viscous deformation can take place at any state of stress.

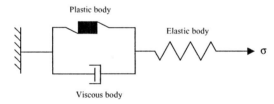

Fig. 5. The combination of a plastic St. Venant body parallel to a viscous Newton body and in series with an elastic Hooke body is called an extended Bingham body. The viscosity of the Newton body is either linear (model 1) or non-linear (model 2 and 3).

The expression for the deviatoric shear stress tensor σ'_{ij} of the extended Bingham body is

$$\sigma'_{ij} = 2G \cdot \varepsilon'_{ij} \qquad \qquad \textit{for } \sigma \leq \sigma_y \qquad (2)$$

$$\sigma'_{ij} = \sigma_y + 2\eta \cdot \dot{\varepsilon}'_{ij} \qquad \qquad \textit{for } \sigma > \sigma_y$$

where G is the shear modulus, ε'_{ij} the deviatoric strain, σ_y the yield strength, η the viscosity and $\dot{\varepsilon}'_{ij}$ the time derivative of the deviatoric strain rate. The viscosity can also be described with the effective viscosity η_{eff} determined from a power law, which is in the presented models a non-linear temperature dependent dislocation creep. As a function of stress the effective viscosity can be expressed as

$$\eta_{eff} = \tfrac{1}{2A} \cdot e^{(Q/RT)} \cdot (\sigma_1 - \sigma_3)^{(1-n)} \tag{3}$$

where A is a material constant, n the stress exponent, Q the activation energy, R the gas constant, T the temperature in Kelvin, and σ_1, σ_3 the largest and the smallest principal stress. The values taken for the power law are listed in Table 1. The density of the upper crust is 2750 kg m^{-3}, 2950 kg m^{-3} for the lower crust and 3150 kg m^{-3} for the upper mantle and the oceanic crust.

Table 1. Parameters for temperature controlled-dislocation creep.

Rock type or mineral	A [Pa^{-n} s^{-1}]	n	Q [k J mol^{-1}]
Granite (dry, lower crust)	$3.16 \cdot 10^{-26}$	3.3	186.3
Granite (wet, lower crust)	$7.94 \cdot 10^{-16}$	1.9	140.6
Diabas (dry, lower crust)	$6.31 \cdot 10^{-20}$	3.05	276.0
Diabas (wet, lower crust)	$1.26 \cdot 10^{-16}$	2.4	212.0
Olivin (dry, upper mantle)	$7.00 \cdot 10^{-14}$	3.0	510.0
Olivin (wet, upper mantle)	$3.98 \cdot 10^{-25}$	4.5	498.0

Data are taken from Carter and Tsenn (1987). A=material constant, n=stress exponent, Q=activation energy.

7. Model Results

Three different models will be presented. Model 1 one uses linear Newton viscosity and focuses on the influence of a variation of the slab pull stresses and different friction coefficients along the faults and the subduction zones. In model 2 and 3 non-linear viscosity is applied. Model 3 also considers the influence of different cohesive strengths for extensional and compressional tectonic regimes.

7.1 Model 1 with Linear Newton Viscosity

In model 1 a viscosity of 10^{21} Pa s for the lower crust and 10^{22} Pa s for the upper mantle is applied. The various model parameter studies revealed that in order to get reasonable displacement rates the friction coefficient has to be reduced to values of 0.2 for the North and East Anatolian fault and the Dead Sea fault and rift system, 0.35 for the Hellenic arc and 0.45 for the Cypriotic arc and the suture zones in the Dinarides and the Apennines. In this case the directions of the velocity field are in relative good agreement with the GPS results ($\pm 21°$ in average), but the magnitudes are smaller and deviate in average by 38% (Fig. 6). The modeled velocities are between 0.6 and 1.2 cm/a while the geodetic observations are up to 3.0 cm/a (McClusky et al. 2000) and geologic data up to 2.5 cm/a for Anatolia along the eastern part of the North Anatolian fault (Barka and Hancock 1984; Barka and Kadinsky-Cade 1988). However, the lateral westward escape along the North Anatolian fault and the south-southwest movement of the Aegean region is reflected in the model. Within the Anatolian-Aegean block areas with different strain rates can be identified. Eastern Anatolia has low strain rates, i.e. a relatively rigid body movement, and the Aegean region has higher strain rates resulting from the dilatation in that region. The model strain rate in the Aegean Sea between $3 \cdot 10^{-15}$ s^{-1} and $5 \cdot 10^{-15}$ s^{-1}) is in good agreement with the seismological determined strain rates from Jackson and McKenzie (1988). In comparison with the geodetic observations a value of 150 MPa for the slab pull stresses (equals $1.5 \cdot 10^{13}$ N m^{-1} of a 100 km thick lithosphere) gives the best result. Parameter studies with higher friction coefficients (0.6-0.9) result in strong coupling along the North Anatolian fault and no westward escape, which is not consistent with geologic or geodetic data. HHigher slab pull stresses (200-300 MPa) give unreasonable high subsidence rates of more than 0.8 cm/a in the Aegean Sea.

A reduction of the viscosity in the lower crust and the upper mantle by one order of magnitude in the Aegean region did not affect the velocity field or the strain rate field at the surface. The effect of a higher density contrast between the two crustal layers and the upper mantle (increase of the density contrast from 200 kg m^{-3} to 250 kg m^{-3}) on the surface deformation pattern is also small. However, assuming no density contrast, i.e. applying a uniform density of 3000 kg m^{-3} for the whole model, reduced the velocity and strain rates in western Anatolia.

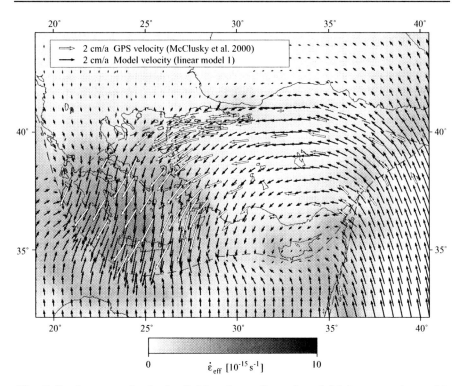

Fig. 6. Strain rate and velocity field at the surface of model 1 in comparison with GPS results. Model 1 has linear Newton viscosity and homogeneous cohesive strength of 100 MPa. Further parameters are: 150 MPa slab pull stresses, friction coefficients of 0.2-0.45 along the faults.

7.2 Model 2 and 3 with Non-linear Viscosity

In model 2 and 3 a temperature controlled dislocation creep is applied. The temperature field was calculated from measured surface heat flow data (European Thermal Atlas; Hurtig 1995) and from data for heat production and thermal conductivity taken from publications of Čermák and Brodi (1995) and Seipold (1995). In model 2 the results show that the non-linear viscosity leads to an amplification of strain rates between $6 \cdot 10^{-15}$ s^{-1} and $8 \cdot 10^{-15}$ s^{-1}). Velocities increase in the Aegean, but decrease in the Anatolian region (Fig. 7). Again, a variation of the effective viscosity has only small effects on the strain rate and the velocity field at the surface.

The indentation of the Arabian plate into Eurasia does not lead to a sufficiently fast westward movement of the Anatolian block. The kinematic energy from the indentation process is partly consumed by compressive deformation processes.

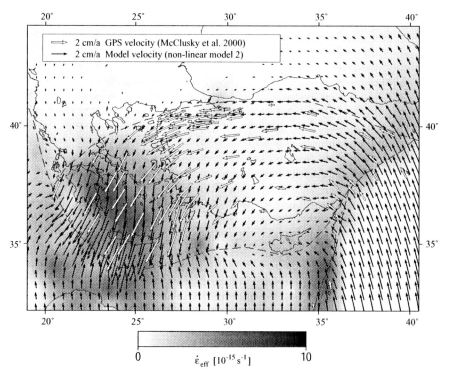

Fig. 7. Strain rate and velocity field at the surface of model 2 in comparison with GPS results. Model 2 has non-linear viscosity and homogeneous cohesive strength of 100 MPa. Further parameters are: 150 MPa slab pull stresses, friction coefficients of 0.2-0.45 along the faults.

In model 3 the effect of variations in the cohesive strength C is investigated. After Brace (1964) the cohesive strength of granite is under compression up to eleven to twelve times higher than under extension. Therefore in the Anatolian region and in the Caucasian mountains, which are, as a first approximation, under compression (Müller et al. 1992), a cohesive strength of 460 MPa is assumed. For the Aegean which is mainly affected by an extensional tectonics, the cohesive strength is set to 40 MPa.

As a result the velocity in the Anatolian region increased and values of 2.0 cm/a are reached (Fig. 8). The strain rates decreased slightly. As a first approximation the Anatolian block behaves as a rigid unit. In the Aegean block the effect of the lowered cohesive strength is smaller and can only be observed in the western and northwestern parts. Most of the changes are

due to the higher yield strength in Anatolia. Calculating the deviation between the GPS results and the velocities from model 3 at the GPS observation sites, the mean deviation is ±10° for the azimuth and ±21% for the
magnitudes (Fig. 9). Large deviations occur in western Anatolia and south
of the Marmara Sea. In the Anatolian block and near the Hellenic arc deviations are low.

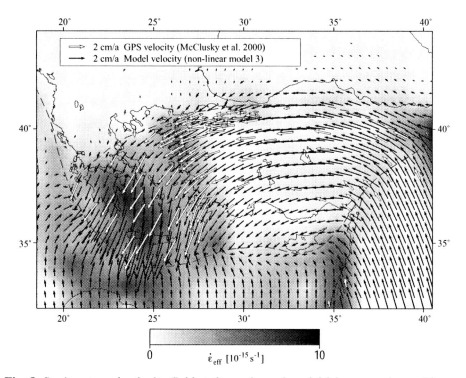

Fig. 8. Strain rate and velocity field at the surface of model 3 in comparison with
GPS results. Model 3 has non-linear viscosity and variable cohesive strength
(40 MPa in the Aegean block, 460 MPa in the Anatolian block and the Caucasian
region). Further parameters are: 150 MPa slab pull stresses, friction coefficients of
0.2-0.45 along the faults.

8. Discussion and Conclusions

Changes of the viscosity (model 1) or of the effective viscosity (model 2
and 3) of the lower crust and the upper mantle had only little effects on the
surface deformation and the main pattern of the velocity field. The surface
velocities are mainly controlled by the yield strength of the upper crust, the

friction coefficient applied to the modeled faults and the magnitudes of the slab pull stresses. The model results support the assumption made for this model that the subduction process north of the Kephalonia fault and east of Crete towards the Cyprus arc has been terminated. Model calculations with slab pull stresses along the Cyprus arc increased the deviation between the observed and the modeled velocity field in southern Anatolia. The models show that the main driving forces for the deformation process in the Eastern Mediterranean are the indentation of the Arabian plate into Eurasia and the retreat of the Hellenic arc subduction zone induced by the slab pull stresses. Both processes are needed for the westward escape of the Anatolian-Aegean block. The indentation provides most of the energy required for the lateral extrusion and the retreat provides the space needed for the escape of the block. These two processes are interweaved.

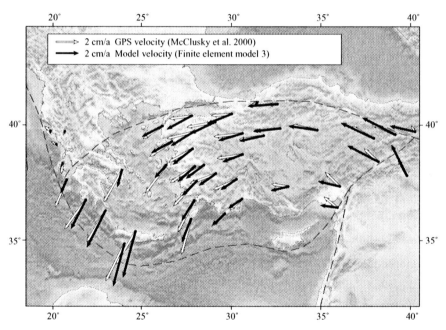

Fig. 9. Velocity vectors at 45 GPS sites from McClusky et al. (2000) (white arrows) versus the results from model 3 (black arrows). Base map with topography and bathymetry.

The model results show that a mantle drag as an additional boundary condition is not necessary as a rule. This does not imply that other additional driving forces do not exists, but they are probably of small importance for the surface processes in the investigated region. In all models the Anatolian region moves almost as a rigid unit which is expressed by low strain rates. In contrast, the Aegean region is strongly deformed through extensional tectonics. This process is induced by the SSW-wards retreat of the Hellenic arc subduction zone. Inside the Anatolian-Aegean block model 3 matches the geodetic observations best, but in the transition zone between western Anatolia and the east Aegean coast line deviations in magnitude and azimuth are significant. These deviations might be explained by four shortcomings of the models: (1) The Strabo-Plini fault system has not been incorporated as a contact surface into the model. Sinistral movements along this fault system could have an important local influence on the velocity field east of Crete. This effect can not account for the large deviations further north. (2) The horizontal pressure gradient due to the varying Moho depth might have been underestimated in the data on which the model geometry is based. Especially in the Aegean Sea and western Anatolia Moho data are relatively sparse. (3) Basal drag caused by mantle flow could act as an additional driving force. (4) Episodic behavior of the upper crustal deformation processes. From the relatively short period of geodetic observations (~10 a) the average values of the velocity field might not reflect the long-term behavior which is represented by the model (0.5 Ma modeling time).

Weak faults (low values for the frictional coefficient) and a strong upper crust in regions with compressional tectonic regime (high cohesive strength) is needed to model the geodetically observed deformation pattern and surface velocity field. The value for frictional sliding along the North Anatolian fault and Hellenic arc is 0.2 which is a third of the proposed value from literature (Kohlstedt et al. 1995). This low value might be caused by a high fluid pressure, from fine grained structure of the rocks within the faults or from frictional heating (Michel and Janssen 1996; Bird and Kong 1994; van de Beukel and Wortel 1988). In the model cohesive strength under compression has to be four times higher than the maximum value given from laboratory investigations (Brace 1964).

The advantage of this three-dimensional approach is that the lithological and rheological layering have been taken into account and slab pull stresses could be introduced without a horizontal parameterization. Even though rheological and lithological layering have only small effects on the surface deformation pattern, the cumulative influence onto the model results can be seen. The modeling can not resolve the local effects caused, e.g. by small-scale fault systems, but it provides valuable boundary conditions which can be used for small- scale models.

Acknowledgments

This work was supported by the Deutsche Forschungsgemeinschaft (DFG, project number Dr 143/5-2). All figures were made with the public domain software GMT (Wessel and Smith 1991). I am very grateful to Blanka Sperner, Hermann Drewes and Zvi Ben-Avraham for their helpful suggestions and improvements on the manuscript.

References

Barka AA, Hancock PI (1984) Neotectonic deformation patterns in the convex northward arc of the North Anatolian fault. In: Dixon JE, Robertson AHF (eds) The Geological Evolution of the Eastern Mediterranean, Geological Society, London, Special Publications 17: 763-777

Barka AA, Kadinsky-Cade C (1988) Strike-slip fault geometry in Turkey and its influence on earthquake activity. Tectonics 7: 663-684

Ben-Avraham Z, Lyakhovsky V, Grasso, M (1995) Simulation of collision zone segmentation in the central Mediterranean. Tectonophysics 243: 57-68

Ben-Avraham Z, Ginzburg A (1990) Displaced Terranes and Crustal Evolution of the Levant and the Eastern Mediterranean. Tectonics 9: 613-622

Ben-Avraham Z, Kempler D, Ginzburg A (1988) Plate convergence in the Cyprian Arc. Tectonophysics 146: 231-240

Bird P, Kong X (1994) Computer simulations of California tectonics confirm very low strength of major faults. Geological Society of America Bulletin 106: 159-174

Brace WF (1964) Brittle fracture of rocks. In: Judd WR (ed) State of Stress in the Earth's crust. Elsevier, New York

Carter NJ, Tsenn MC (1987) Flow properties of continental lithosphere. Tectonophysics 136: 27-63

Cermák V, Bodri L (1995) Three-dimensional deep temperature modeling along the European geotraverse. Tectonophysics 244: 1-11

Cianetti S, Gasperini P, Boccaletti M, Giunchi C (1997) Reproducing the velocity and stress fields in the Aegean region. Geophysical Research Letters 24: 2087-2090

de Jonge M, Wortel MJR, Spakman W (1994) Regional scale tectonic evolution and the seismic velocity structure of the lithosphere and upper mantle: The Mediterranean region. Journal of Geophysical Research 99: 12091-12108

DeMets C, Gordon RG, Argus DF, Stein S (1994) Effect of recent revisions to the geomagnetic reversal time scale on estimates of current plate motions. Geophysical Research Letters 21: 2191-2194

Dewey JF, Helman ML, Turco E, Hutton DHW, Knott SD (1989) Kinematics of the western Mediterranean. In: Coward H, Dietrich D, Park RG (eds) Alpine Tectonics. Blackwell Scientific Publications, Oxford, pp 265-283

Geiss E (1987) A new compilation of crustal thickness data for the Mediterranean area. Annali di Geofisica 5B: 623-630

Heidbach O (2000) Der Mittelmeerraum - Numerische Modellierung der Lithosphären-dynamik im Vergleich mit Ergebnissen aus der Satellitengeodäsie. PhD thesis, Deutsche Geodätische Kommission, Serie C: Dissertationen, Nr. 525, München

Heidbach O, Drewes H (2003) 3-D Finite Element model of major tectonic processes in the Eastern Mediterranean. In: Nieuwland D (ed) New insights in structural interpretation and modeling, Geological Society, London, Special Publications 212: 259-272

Hurtig E. (1995) Temperature and heat-flow density along European transcontinental profiles. Tectonophysics 244: 75-83

Jackson J, Haines J, Holt W (1992) The horizontal velocity field in the deforming Aegean Sea region determined from the moment tensors of earthquakes. Journal of Geophysical Research 97: 17657-17684

Monitoring of Slab Detachment in the Carpathians

Blanka Sperner & the CRC 461 Team

Geophysical Institute, Karlsruhe University, Germany

Abstract

Detachment of descending oceanic lithosphere (slab) is considered to be an important geodynamic process returning lithospheric material into the deeper mantle. On the geological time scale the detachment process itself is a short-term procedure. Thus it can be studied only at a few localities on Earth today. The Vrancea region in the SE-Carpathians is one of these rare places. In the SE-Carpathians, Miocene subduction was followed by continental collision accompanied by slab steepening (rollback). At the present time the slab is in a nearly vertical position and we interpret its strong intermediate-depth seismicity (70-180 km depth) to be triggered by slab pull (vertical extension axes from focal mechanism solutions). Our results from seismic tomography revealed a high-velocity body beneath the SE-Carpathians extending to a depth of at least 350 km and showing a SW-NE orientation. Seismicity is restricted to the north-eastern part of the high-velocity body. The SW-part is aseismic and thus probably already detached from the overlying crust. Detachment of the NE-part is likely to happen at the moment or in near future at a depth range of 40 to 70 km where a zone of low seismicity is associated with a low-velocity subcrustal layer detected in refraction seismic profiles. The seismicity is located at the lower surface of the descending slab and thus can be interpreted as the lower plane of a double seismic zone where earthquakes are triggered be dehydration reactions. The upper plane, which is not observed in Vrancea, might be missing due to different reaction times, water contents, or other parameters varying for the two different slab regions. We developed a new model for the regional tectonic evolution during the last

15 million years taking into consideration subduction zone retreat, slab rollback and finally detachment. In this model the geometry of the plate boundaries define the direction of subduction retreat and thus the present-day NE-SW orientation of the slab. The subdivision of the slab in a seismically active NE-part and an aseismic SW-part is explained by slab detachment of the SW-part. The NE-part is still (partly) attached so that slab-pull induced stress accumulations lead to the intermediate-depth seismicity. We tested different scenarios for the tectonic evolution of the SE-Carpathians by kinematic gravity modelling and found a best-fit of modelled and measured gravity anomaly data for slab delamination and a still attached slab in the Vrancea region.

1. Introduction

The frequent occurrence of strong earthquakes ($M_w > 6.8$) near Bucharest, the capital city of Romania, initiated intense research activity in this southeastern part of the Carpathian arc (e.g. Wenzel et al. 1999). In 1996 and 1997 two closely collaborating research projects there established, the "Romanian Group for Strong Earthquakes (RGVE)" under the umbrella of the Romanian Academy (Bucharest) and the Collaborative Research Center (CRC) 461 "Strong Earthquakes: A Challenge for Geosciences and Civil Engineering" at Karlsruhe University (http://www-sfb461.physik. uni-karlsruhe.de/) as a project of the Deutsche Forschungsgemeinschaft (DFG). In both groups scientists from different fields (geology, geophysics, civil engineering, operation research) make a multidisciplinary attempt towards earthquake mitigation. Key objectives of the joint research activities are (i) the understanding of the tectonic processes that cause the strong intermediate depth seismicity, (ii) the development of realistic models and predictions of ground motion, (iii) the prognosis of potential damage in case of a strong earthquake, and (iv) risk reduction by appropriate civil engineering concepts. In this paper we focus on the first topic, i.e. the geodynamic evolution of the SE-Carpathians and the present-day situation in the seismically active region (Vrancea). We reveal new insights into the processes active in Vrancea by the joint interpretation of data from different fields and different depth levels, namely from the surface (gravity data), the crust (refraction seismics) and the mantle (seismic tomography). We present preliminary results from seismic tomography, discuss possible earthquake triggering mechanisms, and integrate the data into a model for the present slab situation and the geodynamic evolution of the region

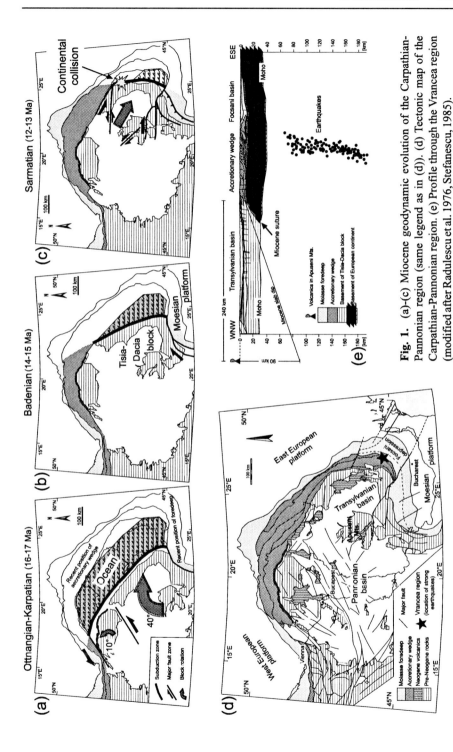

Fig. 1. (a)–(c) Miocene geodynamic evolution of the Carpathian-Pannonian region (same legend as in (d)). (d) Tectonic map of the Carpathian-Pannonian region. (e) Profile through the Vrancea region (modified after Radulescu et al. 1976, Stefanescu, 1985).

during the last 15 Ma. Finally, we test this model and alternative ideas by means of kinematic gravity modelling.

2. Tectonic Overview

The Neogene evolution of the Carpathians is mainly driven by the NE- and later E-ward retreat of a SW-, later W-dipping subduction zone (Fig. 1a-c) (Sperner et al. 2001). The upper plate consisted of two blocks moving and rotating into different directions (Fig. 1a) (e.g. Balla 1985, Csontos 1995). Subduction retreat pulled them into an oceanic embayment, the last remnant of the Alpine Tethys south of the European continent (Stampfli and Borel 2002). When the European continent started to enter the subduction zone, the buoyancy forces of the thick continental crust exceeded the slab pull forces and convergence stopped after an only short period of continental thrusting. The age of the youngest thrusting in the fold-and-thrust belt of the accretionary wedge decreases from 13 Ma in the northern part to 10 Ma in the south-eastern part (Jiricek 1979). This indicates that collision first took place in the northern part of the Carpathian arc while subduction was still going on for a short time in the eastern part (Fig. 1b, c) (Sperner et al. 1999). The arcuate shape of the Carpathians (Fig. 1d) was predefined by the shape of the oceanic embayment and represents a primary arc, i.e. it already developed during the initial formation of the mountain belt (Zweigel et al. 1998). In contrast, secondary arcs are originally straight and are bent later.

Today the subducted lithosphere beneath Vrancea is in a sub-vertical position as indicated by the distribution of intermediate-depth earthquakes (Fig. 1e) and by the results of seismic tomography (Sperner et al. 2001, Wenzel et al. 1998, Wortel and Spakman 2000). The Miocene dip angle of the subducting slab can be determined using the fact that at depths of 80-100 km dewatering of the down-going slab induces melting of the mantle wedge above the slab and thus volcanism at the surface. Subduction-related calc-alkaline volcanism of Miocene age (9.3-14.7 Ma) (Pécskay et al. 1995) is known from the Apuseni Mountains, at a distance of 240 km from the plate boundary (Fig. 1e). The depth of the dewatering slab (90 km) and the distance volcanic rocks - plate boundary (240 km) define a dip angle of *c.* 20° (Fig. 1e). After convergence ended in Vrancea ~10 Ma ago, the initially flat slab began to steepen to its present-day sub-vertical position.

3. Geophysical Data

In 1999 a tomographic experiment with 120 seismic stations was carried out by the CALIXTO (**C**arpathian **A**rc **Li**thosphere **X-To**mography) group consisting of researchers from Bucharest University, NIEP Bucharest-Magurele, EOST Strasbourg, ETH Zürich, IRRS Milan and Karlsruhe University (Martin et al. 2001). The station distance ranged from 15-20 km (Vrancea region) to 25-30 km (outer margins of the network) covering a region of about 350 km in diameter. During the 6 months of the field experiment 160 local events with magnitude $M_l \geq 2.0$ and 450 teleseismic events with magnitude $M_b \geq 5.0$ were recorded. First preliminary results were achieved through an inversion of the teleseismic data with the ACH method (Aki et al. 1977, Evans and Achauer 1993). Crustal complexity is responsible for a smearing of these results within the upper 70 km. Varying crustal thickness and young deep basins like the Focsani foredeep basin with its 9 km of Neogene and Quaternary sediments have to be taken into account in the next step of data analysis. But nevertheless the results for the deeper parts are reliable enough to fix the general shape of the velocity anomalies. Due to the crustal smearing there might be an error in position of approximately one block size (*c.* 30 km).

Data inversion reveals a high-velocity body with maximum P-wave velocity perturbations of +3.0 to +3.5% in comparison with the background model (Fig. 2) (Martin et al. 2001). We interpret this high-velocity body as the descending lithospheric slab. It incorporates nearly all intermediate-depth earthquakes and reaches a depth of at least 350 km which is in good agreement with results of previous regional seismic tomography studies (e.g. Wortel and Spakman 2000). Our images show that the high-velocity body has a larger lateral extent than the seismogenic volume, with the SW-part of the slab being aseismic (Fig. 2a).

For the above mentioned reason (smearing of the results due to crustal features), the upper limit of the high-velocity body can at the moment not be determined from seismic tomography. Indications on its upper limit come from refraction seismic data which suggest a low-velocity zone at a depth of 47 to 55 km beneath the Vrancea region (Hauser et al. 2001). This zone coincides with the seismic gap at depths of 40 to 70 km.

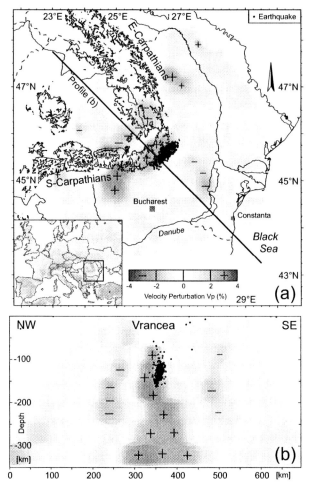

Fig. 2. Results of seismic tomography in the SE-Carpathians. (a) Horizontal section together with earthquakes at 110-150 km depth. (b) Vertical NW-SE section. Earthquakes from a distance up to 10 km were projected into the profile.

4. Seismicity

At the present stage of data analysis, intermediate-depth seismicity in the Vrancea region is concentrated near the south-eastern margin of the north-eastern part of the high-velocity body (Fig. 2) which represented the lower surface of the slab during subduction. Both the intermediate-depth seismicity and the location of the high-velocity body can only be determined within certain error margins. The boundaries of the high-velocity body might shift by up to 30 km (see above) and the earthquake hypocenters are determined with an accuracy of 10 km (K. Bonjer, pers. comm.). Both errors can, in the worst case, sum up to 40 km variance. A (maximum) shift by 40 km would place the seismicity either outside the SE margin or close to the NW margin of the slab, but as long as no better analyses are available we base our interpretation and discussion on the presently identified position at the SE margin of the slab.

The position at the lower surface of the slab is different from the geometry of Wadati-Benioff zones of actively subducting lithosphere where earthquakes are concentrated near the upper surface of the slab (Benioff 1949, Wadati 1935). In Wadati-Benioff zones the triggering of intermediate-depth earthquakes is controlled by the various forces that drive subduction (gravity) and resist subduction (friction with the overriding plate) as well as by plate flexure and thermal effects such as expansion and metamorphic densification (Spence 1987). However, when subduction and collision are terminated, friction and plate flexure become irrelevant and gravity dominates the force balance. This is reflected in the orientation of the stress axes of the intermediate-depth earthquakes which are characterized by down-dip (i.e. vertical) extension as determined from focal-mechanism solutions of individual events (e.g. Constantinescu and Enescu 1964, Oncescu and Bonjer 1997) as well as from the inversion of all earthquake data together (Oncescu 1987, Plenefisch 1996).

As an additional process dehydration reactions might play a significant role in triggering earthquakes in the Vrancea region. The water released through dehydration increases the pore pressure, so that the earthquakes, which are primarily induced by slab pull, can be triggered at lower differential stresses. The best-know dehydration reaction is the gabbro-eclogite transformation in the crustal part of subducting oceanic lithosphere (Kirby et al. 1996), i.e. at the upper slab surface. In Vrancea the earthquakes are located at the lower slab surface and thus might be interpreted as being identical to the lower plane earthquakes of a double seismic zone. Peacock (2001) explains these earthquakes in the subducting oceanic mantle by dehydration embrittlement due to the reaction of antigorite (serpentine) to forsterite + enstatite + H_2O. This reaction occurs at 50-200 km depths, thus coinciding with the depth range of the Vrancea earthquakes. When we interpret the Vrancea earthquakes as the lower plane earthquakes of a double seismic zone, we need a reason for the lack of the upper plane earthquakes. Explanations might be found (a) in different reaction times for the two dehydration processes, (b) in different water contents of the different slab regions, or (c) in different water release conditions. Water release might be easier at the upper surface of the slab where the hot and low-viscosity mantle wedge absorbs the ascending hydrous phases. In contrast, the lower plane is overlain by cold and high-viscosity slab material which might hinder the migration of the releasing water. As long as subduction provides a permanent water supply from the surface, these differences might be negligible. But as soon as continental collision stops the water supply, a faster dehydration of the upper plane might lead to a decrease and finally the expiration of seismicity.

Another explanation for the location of seismicity at the lower slab surface is discussed by Ismail-Zadeh et al. (this volume). Their analytical modelling of the maximum shear stress takes into account the specific situation in Vrancea and also includes different mantle viscosities on both sides of the slab considering the different evolution of the old and thus cold European craton on the one side and the young and warm Tisia-Dacia block on the other side. This configuration leads to a concentration of the maximum shear stress at the lower slab surface (Ismail-Zadeh et al., this volume), in contrast to an active subduction zone where the maximum shear stress is located at the upper surface of the slab (McKenzie 1969). The depth interval of the shear stress maximum is consistent with the depth range of the intermediate-depth seismicity in Vrancea.

5. Geodynamic Model

The new tomography image requires modifications of the previous geodynamic models and slab geometries. So far an eastward retreat of the subduction zone was assumed for the final phase of subduction (e.g. Csontos et al. 1992, Fodor et al. 1999, Sperner et al. 2001). The now observed NE-SW orientation of the slab points to a change of the retreat direction from E- to SE-wards during the end of subduction (Fig. 3). The reason is an oblique collision of the overriding block (Tisia-Dacia block) with the European foreland caused by different orientations of the converging continental margins. The eastern margin of the Tisia-Dacia block had a N-S orientation, while the European foreland was NW-SE oriented (Fig. 1c). Thus, continental collision first occurred in the northern part of the Eastern Carpathians and as a consequence horizontal movements were blocked in this region (Fig. 1a-c). The still ongoing convergence in the southern part now took place between two differently oriented lateral boundaries, the E-W oriented northern margin of the Moesian platform and the NNW-SSE

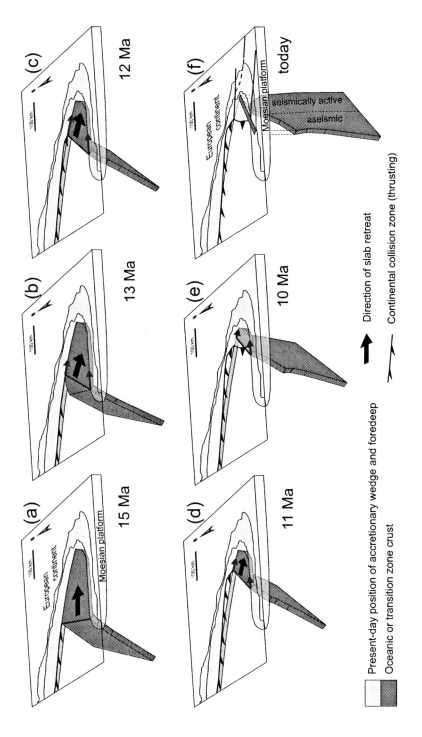

Fig. 3. Model for the plate-tectonic evolution of the SE-Carpathians from middle Miocene until today. The overriding block had been removed for better view on the slab. Retreat direction changes from E- to SE-wards, so that the slab is pulled beneath the Moesian platform (Fig. 1).

oriented western margin of the European continent. Along both margins the subducting lithosphere was torn off during subduction retreat (small arrows in Fig. 3) and the retreat direction followed the bisecting line between the two margins towards SE. As a consequence of the change in retreat direction, a part of the already subducted and detached slab was pulled beneath the northern margin of the Moesian platform (Fig. 3d-g). This already detached part forms today the aseismic south-western portion of the descending lithosphere. The still attached NE-part of the slab experiences stresses caused by the gravitational sinking of the heavy subducted lithosphere. The related seismicity is restricted to the middle depth interval of the slab between 70 and 180 km. The seismic gap between 40 and 70 km depth, which coincides with the low-velocity zone between 47 and 55 km in refraction seismic data, is interpreted as a zone of weakened mantle or lower crustal material where slab detachment started to take place.

6. Kinematic Gravity Modelling

We used kinematic gravity modelling to test different scenarios for the tectonic evolution of the SE-Carpathians. Kinematic gravity modelling differs from standard gravity modelling in two ways: it includes the tectonic evolution of the study area and it considers isostatic adjustments (Lillie 1991, Lillie et al. 1994). We paid special attention to the evolution of the slab during and after subduction and its influence on vertical surface movements (due to isostatic balancing) as well as on the large-scale pattern of the gravity anomalies (Bouguer and free air). In our model middle Miocene oceanic subduction resulted in the evolution of a lithospheric root beneath the Tisia-Dacia block. Due to the higher density of this root (in comparison with the surrounding asthenosphere) the Tisia-Dacia block was pulled downward so that the Transylvanian basin started to develop (Fig. 4a). After continental collision (~10 Ma) steepening of the slab shifted the lithospheric root trenchward (Fig. 4b). As a result of this shift, the depocentre of the Transylvanian basin migrated into the same direction (south-eastward; Fig. 4c), while the north-western part of the Transylvanian basin started to uplift due to the removal of the downward-pulling lithospheric root (Fig. 4b).

The modelled gravity anomalies for both stages (Fig. 4a, b) do not match the measured data (Fig. 4d), so that we assume an additional process following the collisional and steepening stages. We tested two options, slab detachment and mantle delamination along the Moho. Slab detachment results in flat and wide gravity anomalies, inconsistent with the

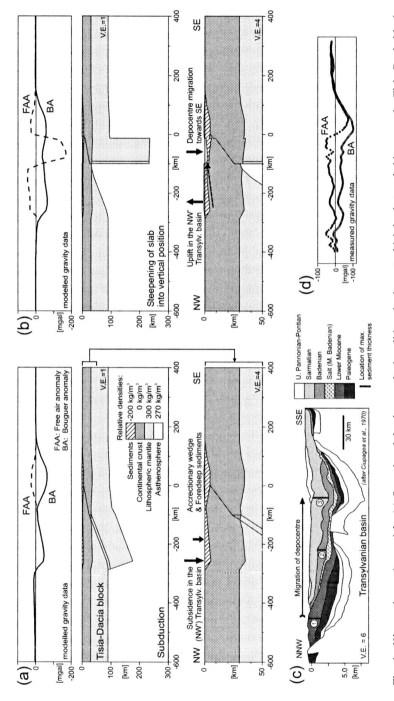

Fig. 4. Kinematic gravity models. (a) Oceanic subduction results in a lithospheric root which leads to subsidence on the Tisia-Dacia block (Transylvanian basin). (b) Steepening of the slab (after continental collision) shifts the lithospheric root and thus the depocentre of the Transylvanian basin towards the (former) trench, i.e. towards the SE (c). The modelled gravity anomalies of both stages are not consistent with the measured gravity data in the SE-Carpathians (d).

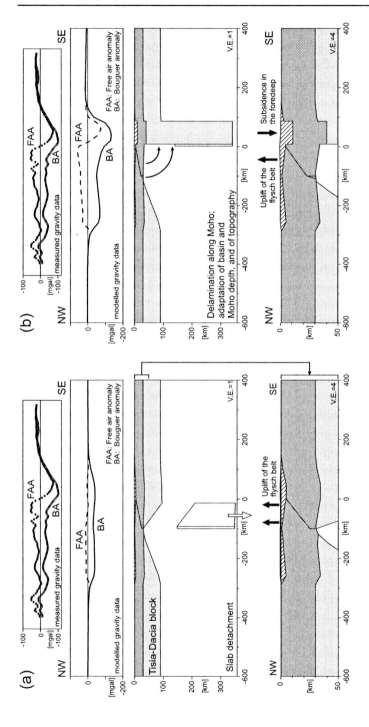

Fig. 5. Kinematic gravity models. (a) Slab detachment releases the overlying lithosphere from the downward pulling slab and thus results in uplift. The modelled gravity data do not coincide with the measured ones. (b) Mantle delamination along the Moho releases parts of the overlying crust and results in even more uplift as in the detachment model because of the removal of the whole lithospheric mantle. At the same time slab pull causes subsidence in the foredeep. The modelled gravity data fit well with the measured ones, if topography as well as Moho and basin depth are adapted to measured data e.g. from seismics.

measured data which show pronounced local minima (Fig. 5a). Mantle de-lamination, as postulated for the SE-Carpathians by Gîrbacea and Frisch (1998), concentrates slab pull beneath the foredeep leading to a deep basin accompanied by a deep Moho and thus a large negative Bouguer anomaly (Fig. 5b). We adopted our model geometry, i.e. basin and Moho depth as well as topography, to the real data as known e.g. from refraction seismics (Hauser et al. 2001). The modelled gravity anomalies coincide in the main with the measured data (Fig. 5b top). Also the modelled vertical move-ments (Fig. 5b bottom) find their equivalent in nature with basin subsi-dence and normal faulting being still active in the foredeep (Matenco et al. 2002) while uplift started in the SE-Carpathian flysch belt at *c.* 5 Ma (Sanders et al. 1999).

7. Conclusions

The analysis of new geophysical data from the Vrancea region in the SE-Carpathians along with the geological knowledge about the tectonic evolu-tion of the Carpathian arc revealed new insights into the geodynamic evo-lution of the region and the fate of the subducted lithosphere. Subduction retreat was the main driving mechanism for the tectonic movements during Miocene times. The retreat direction was in its last phase defined by the laterally bounding continental margins, in the beginning solely by the E-trending margin of the Moesian platform in the south and later addition-ally by the NNW-trending margin of the European continent in the NE. This change in the boundary conditions caused a change of the retreat di-rection from E- to SE-wards and a rotation of the slab into its present-day SW-NE orientation as seen in seismic tomography. Intermediate-depth seismicity occurs only in one part of the slab while the other part is aseis-mic. We conclude that the aseismic part of the slab is already detached from the overlying crust while the other part is still attached. Stresses caused by the gravitationally sinking slab can only accumulate in the at-tached part of the slab. Additionally, dehydration reactions might influence the stress field through an increased pore pressure which promotes brittle failure. A weak zone probably at 40-70 km depth mechanically decouples the slab from the overlying crust representing the region of past and pre-sent delamination of the lithospheric mantle (layer-parallel) or of present (or future?) slab detachment (layer-orthogonal). On the basis of our data we conclude that the detachment of the oceanic lithosphere beneath the SE-Carpathians is an ongoing process in its final stage. Thus, the Vrancea region in the SE-Carpathians is one of the rare places on Earth (if not the only one) where the final stage of slab detachment can be studied today.

Acknowledgements

We acknowledge helpful and critical reviews by Giuliano F. Panza and Birgit Müller. The Collaborative Research Center (CRC) 461 "Strong Earthquakes: A Challenge for Geosciences and Civil Engineering" is funded by the Deutsche Forschungsgemeinschaft (German Science Foundation) and supported by the State of Baden-Württemberg and the University of Karlsruhe.

References

Aki K, Christofferson A, Husebye ES (1977) Determination of the three dimensional seismic structure of the lithosphere. Journal of Geophysical Research 82: 277-296

Balla Z (1985) The Carpathian Loop and the Pannonian Basin: a kinematic analysis. Geophys. Transaction 30: 313-353

Benioff H (1949) Seismic evidence for the fault original of oceanic deeps. Bulletin of the Geological Society of America 60: 1837-1856

Constantinescu L, Enescu D (1964) Fault-plane solutions for some Roumanian earthquakes and their seismotectonic implication. Journal of Geophysical Research 69: 667-674

Csontos L (1995) Tertiary tectonic evolution of the Intra-Carpathian area: a review. Acta Vulcanologica 7: 1-13

Csontos L, Nagymarosy A, Horvath F, Kovac M (1992) Tertiary evolution of the intra-Carpathian area; a model. Tectonophysics 208: 221-241

Evans JR, Achauer U (1993) Teleseismic velocity tomography using the ACH-method: Theory and application to continental-scale studies. In: Iyer KM, Hirahara K (eds) Seismic Tomography; Theory and Practice. Chapman & Hall, London, pp 319-360

Fodor L, Csontos L, Bada G, Györfi I, Benkovics L (1999) Tertiary tectonic evolution of the Pannonian Basin system and neighbouring orogens: a new synthesis of paleostress data. Geological Society, London, Special Publications 156: 295-334

Gîrbacea R, Frisch W (1998) Slab in the wrong place; lower lithospheric mantle delamination in the last stage of the eastern Carpathian subduction retreat. Geology 26: 611-614

Hauser F, Raileanu V, Fielitz W, Bala A, Prodehl C, Polonic G, Schulze A (2001) VRANCEA99 - the crustal structure beneath the southeastern Carpathians and the Moesian Platform from a seismic refraction profile in Romania. Tectonophysics 340: 233-256

Ismail-Zadeh AT, Müller B, Wenzel F (2003) Modelling of descending slab evolution beneath the SE-Carpathians: implications for seismicity. In: Wenzel F (ed) Challenges for Earth Sciences in the 21st Century. Springer, Berlin, Heidelberg, this volume

Jiricek R (1979) Tectonic development of the Carpathian arc in the Oligocene and Neogene. In: Mahel M (ed) Tectonic profiles through the Western Carpathians. Geol. Inst. Dionyz Stur, Bratislava, pp 205-214

Kirby SH, Engdahl ER, Denlinger R (1996) Intermediate-depth intraslab earthquakes and arc volcanism as physical expressions of crustal and uppermost mantle metamorphism in subducting slabs. In: Bebout GE, Scholl DW, Kirby SH, Platt JP (eds) Subduction: Top to bottom. American Geophysical Union, Washington, pp 195-214

Lillie RJ (1991) Evolution of gravity anomalies across collisional mountain belts: clues to the amount of continental convergence and underthrusting. Tectonics 10: 672-687

Lillie RJ, Bielik M, Babuska V, Plomerova J (1994) Gravity modelling of the lithosphere in the Eastern Alpine-Western Carpathian-Pannonian Basin region. Tectonophysics 231: 215-235

Martin M, Achauer U, Kissling E, Mocanu V, Musacchio G, Radulian M, Wenzel F, CALIXTO working group (2001) First results from the tomographic experiment CALIXTO '99 in Romania. 26th General Assembly of the European Geophysical Society (EGS), 25-30 March 2001, Nice, France 2001, Geophysical Research Abstracts:

Matenco LC, Bertotti G, Cloetingh SAPL, Schmid SM (2002) Neotectonics of the SE Carpathians foreland: Constraints from kinematic, geomorphological and remote sensing studies. 27th General Assembly of the European Geophysical Society (EGS), 21-26 April 2002, Nice, France 2002, SE4.05-071

McKenzie DP (1969) Speculations on the consequences and causes of plate motions. Geophysical Journal of the Royal Astronomical Society 18: 1-32

Oncescu MC (1987) On the stress tensor in Vrancea region. Journal of Geophysics 62: 62-65

Oncescu MC, Bonjer K-P (1997) A note on the depth recurrence and strain release of large Vrancea earthquakes. Tectonophysics 272: 291-302

Peacock SM (2001) Are the lower planes of double seismic zones caused by serpentine dehydration in subducting oceanic mantle? Geology 29: 299-302

Pécskay Z, Edelstein O, Seghedi I, Szakács A, Kovacs M, Crihan M, Bernad A (1995) K-Ar datings of Neogene-Quaternary calc-alkaline volcanic rocks in Romania. Acta Vulcanologica 7: 53-61

Plenefisch T (1996) Untersuchungen des Spannungsfeldes im Bereich des Rhein-
grabens mittels der Inversion von Herdflächenlösungen und der Abschätzung der bruchspezifischen Reibungsparameter. Ph.D. thesis, Karlsruhe
University.

Radulescu DP, Cornea I, Sandulescu M, Constantinescu P, Radulescu F, Pompi-
lian A (1976) Structure de la croute terrestre en Roumanie - essai d'inter-
prétation des études seismiques profunds. Anural Institutului de Geologie
si Geofizica 50: 5-36

Sanders CAE, Andriessen PAM, Cloetingh SAPL (1999) Life cycle of the East
Carpathian orogen: Erosion history of a doubly vergent critical wedge assessed by fission track thermochronology. Journal of Geophysical Research 104: 29,095-29,112

Spence W (1987) Slab pull and the seismotectonics of subducting lithosphere. Reviews of Geophysics 25: 55-69

Sperner B, Ratschbacher L, Zweigel P, Moser F, Hettel S, Gîrbacea R, Wenzel F
(1999) Lateral extrusion, slab-break-off and subduction retreat: the Oli-
gocene-Recent collision-subduction transition in the Alps and Carpathians. Penrose Conference "Subduction to Strike-Slip Transitions on Plate
Boundaries" (available on-line at http://www.uncwil.edu/people/
grindlayn/penrose.html), Puerto Plata, Domin. Rep. 1999, 103-104

Sperner B, Lorenz FP, Bonjer K-P, Hettel S, Müller B, Wenzel F (2001) Slab
break-off - abrupt cut or gradual detachment? New insights from the
Vrancea Region (SE Carpathians, Romania). Terra Nova 13: 172-179

Stampfli GM, Borel GD (2002) A plate tectonic model for the Paleozoic and
Mesozoic constrained by dynamic plate boundaries and restored synthetic
oceanic isochrons. Earth and Planetary Science Letters 196: 17-33

Stefanescu M (1985) Geologic profile A-14, 1:200,000. Geological Institute, Bucharest

Wadati K (1935) On the activity of deep-focus earthquakes in the Japanese Islands
and neighborhoods. Geophysical Magazine 8: 305-325

Wenzel F, Achauer U, Enescu E, Kissling E, Russo R, Mocanu V, Musacchio G
(1998) Detailed look at final stage of plate break-off is target of study in
Romania. EOS Transactions 79: 589-594

Wenzel F, Lungu D, Novak O (1999) Vrancea Earthquakes: Tectonics, Hazard
and Risk Mitigation. Kluwer Academic Publishers, Dordrecht Boston
London

Wortel MJR, Spakman W (2000) Subduction and slab detachment in the Mediterranean-Carpathian region. Science 290: 1910-1917

Zweigel P, Ratschbacher L, Frisch W (1998) Kinematics of an arcuate fold-thrust
belt: the southern Eastern Carpathians (Romania). Tectonophysics 297:
177-207

Modelling of Descending Slab Evolution Beneath the SE-Carpathians: Implications for Seismicity

A. Ismail-Zadeh[1], B.Müller, and F. Wenzel

Geophysikalisches Institut, Universität Karlsruhe, Germany

Abstract

Recent findings from regional seismic tomography and refraction studies and from GPS studies on vertical movements together with extremely high intermediate-depth seismicity in the Vrancea region (Romania) point towards the interpretation that the lithospheric slab, descending beneath the SE-Carpathians, approaches a stage of break-off or even it is already delaminated from the crust. To understand processes of stress generation due to the descending slab, we analyse tectonic stress, induced by the slab sinking in the mantle, by means of analytical and numerical modelling. We find that the maximum shear stress migrates from the upper surface of the slab to its lower surface in the course of changes in slab dynamics from its active subduction through roll-back movements to sinking solely due to gravity. The changes in stress distribution can explain the location of hypocentres of Vrancea events at the side of the slab adjacent to the Eastern European craton. To analyse a process of slab delamination, we develop a two-dimensional thermomechanical finite-element model of a slab sinking in the mantle due to gravity and overlain by the continental crust. The model predicts lateral compression in the slab as inferred from the stress axes of earthquakes and its thinning and necking. The maximum stress occurs in the depth range of 80 km to 200 km, and the minimum stress falls into the depth range of 40 km to 80 km. The area of maximum shear stress coincides with the region of high seismicity, and minimum shear stress in

---

[1]   On leave from the International Institute of Earthquake Prediction Theory and Mathematical Geophysics, Russian Academy of Sciences, Moscow.

the model is associated with the lower viscosity zone. Uplift of the crust starts before the slab detachment. Just after the detachment the tectonic stress in the slab is still high enough to lead to a seismic activity, and the stress decreases significantly in several million years.

1. Introduction: Vrancea Seismicity and Geodynamics

Repeated intermediate-depth large earthquakes of the SE-Carpathians (Vrancea) cause destructions in Bucharest and shake the central and eastern European cities at distances of several hundred kilometres away from the hypocentres of the events. The earthquake-prone Vrancea region is situated at the bend of the SE-Carpathians and bounded on the north and north-east by the Eastern European craton (EEC), on the east and south by the Moesian platform (MP), and on the west by the Transylvanian basin (TB).

The epicentres of mantle earthquakes in the Vrancea region are concentrated within a very small area (Fig. 1), and the distribution of the epicentres is much denser than that of intermediate-depth events in other intracontinental regions. The projection of the foci on the NW-SE vertical plane across the bend of the Eastern Carpathians (section AB in Fig.1) shows a seismogenic volume about 100 km long, about 40 km wide, and extending to a depth of about 180 km. The body is interpreted as a lithospheric slab descending in the mantle. Beyond this depth the seismicity ends suddenly: a seismic event beneath 180 km represents an exception. A seismic gap at depths of 40-70 km led to the assumption that the lithospheric slab was already detached. According to the historical catalogue of Vrancea events, large intermediate-depth shocks with magnitudes $M_W>6.5$ occur three to five times per century. In the XXth century, large events at depths of 70 to 170 km occurred in 1940 with moment magnitude $M_W=7.7$, in 1977 $M_W=7.4$, in 1986 $M_W=7.1$, and in 1990 $M_W=6.9$ (Oncescu and Bonjer 1997).

The 1940 earthquake gave rise for the development of a number of geodynamic models for this region. Gutenberg and Richter (1954) drew attention to the Vrancea region as a place of remarkable intermediate depth seismicity. Later McKenzie (1972) suggested this seismicity to be associated with a relic slab sinking in the mantle and now overlain by continental crust. The 1977 disastrous earthquake and later the 1986 and 1990 earthquakes brought again up the discussion about the nature of the earthquakes. The Vrancea region was considered (Fuchs et al. 1979) as a place

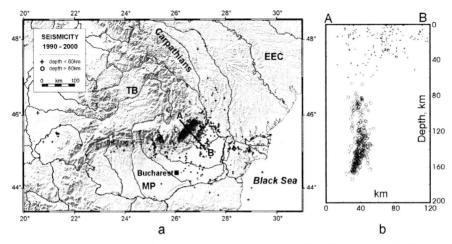

a   b

Fig. 1. Location map of observed seismicity in Romania with magnitude $M_W \geq 3$ (after Sperner et al. 2001). (a) Epicentres of Vrancea earthquakes. (b) Hypocentres of the same earthquakes projected onto the vertical plane AB along the NW-SE direction. TB, Transylvanian basin; EEC, Eastern European craton; MP, Moesian platform.

where the sinking slab was already detached from the continental crust. Oncescu (1984) proposed that the intermediate-depth events are generated in a zone that separates the sinking slab from the neighbouring immobile part of the lithosphere rather than in the sinking slab itself. Linzer (1996) explained the nearly vertical position of the Vrancea slab as the final roll-back stage of a small fragment of oceanic lithosphere. Gibracea and Frisch (1998) assumed that the break-off, affecting only the crustal portion of the slab, was followed by the horizontal delamination of its lower portion. Sperner et al. (2001) suggested a model of Miocene subduction of oceanic lithosphere beneath the Carpathian arc and subsequent gentle continental collision, which transported cold and dense lithospheric material into the mantle. Recent investigations within the framework of the SFB 461 (Collaborative Research Centre: Strong Earthquakes) resulted in modern data sets from seismic refraction techniques, tomography, and geodesy which led to a refinement of the geodynamic model of the Vrancea region (Sperner, this volume).

The active subduction ceased about 10 Ma ago (Wenzel et al. 1998). Subsequently, the initial flat subduction began to steepen to its present-day nearly vertical orientation. Now the cold slab (hence denser than the surrounding mantle) beneath the Vrancea region sinks due to gravity. The hydrostatic buoyancy forces help the slab to subduct, but viscous and frictional forces resist the descent. At intermediate depths these forces produce an internal stress, and earthquakes occur in response to this stress. Ismail-Zadeh et al. (2000) showed that the maximum shear stress in a descending slab accumulates in the depth range of 70 km to 160 km in a very narrow area and the depth distribution of the annual average seismic energy released in earthquakes has a shape similar to that of the depth distribution of the stress magnitude in the slab.

One question to be addressed in this paper concentrates on the pattern of earthquake hypocentres within the slab. Seismic tomographic studies revealed a body of high P-wave velocities beneath the Vrancea region, which was interpreted as a slab descending in the mantle (Wenzel et al. 1998a,b; Bijwaard and Spakman 2000). Its dimensions exceed the seismogenic volume by far.

In 1999 an international tomographic experiment with 120 seismic stations was realised in SE-Romania (Martin et al. 2001). During the field experiment 160 local events with magnitude $M_l \geq 2.0$ and 450 teleseismic events with magnitude $M_b \geq 5.0$ were recorded. The station distance ranged from 15-20 km (the Vrancea region) to 25-30 km (outer margins of the network) covering a region of about 350 km in diameter. First preliminary results were achieved through an inversion of the teleseismic data. Data inversion reveals a high-velocity body with maximum P-wave velocity perturbations of +3.5% in comparison with the background model (see Fig. 2). This high-velocity body is interpreted as the descending lithospheric slab. It reaches a depth of at least 350 km which is in good agreement with results of previous low-resolution seismic tomography studies (e.g., Wortel and Spakman 2000).

The high-resolution seismic tomographic image of the body (Fig. 2; Martin et al. 2001) shows that Vrancea intermediate-depth earthquakes are located at the opposite side of the slab (or lower surface of the descending slab) as compared to zones of active subduction, where seismicity is associated with the upper surface of subducting lithosphere.

Another question is whether the Vrancea slab is still attached to the crust or has been already detached. At the present time the upper limit of the high-velocity body cannot be defined even from the high-resolution seismic tomography. Further information on its upper limit comes from the model based on seismic refraction data: about 10 km high low-velocity zone is detected at a depth of about 50 km beneath the Vrancea region (Hauser et al. 2001). This zone coincides with the seismic gap at same depths and can be interpreted as a zone of weakened mantle (or lower crustal) material.

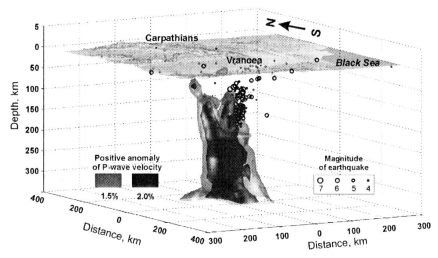

Fig. 2. Seismic tomographic image of the Vrancea slab and hypocentres of intermediate-depth seismicity (based on Martin et al. 2001).

A GPS network operates in the SE-Carpathians since 1997. Recent geodetic measurement revealed relative uplift rates of at least 10 mm a^{-1} (mean uplift rate of 22 mm a^{-1} with an average confidence range of 13.4 mm a^{-1}) in the Vrancea region (Dinter et al. 2001). A possible explanaion of the uplift is that the descending slab is decoupled (or still decoupling) from the overlying crust. Hence the load of the slab decreases, and as a result the released crust starts to uplift.

A principal objective of this contribution is to understand the process of stress generation beneath the Vrancea region by means of analytical and numerical models. We analyse the evolution of shear stress induced by the descending slab from active subduction of the lithosphere to its sinking only due to gravity forces. An analytical model of corner flow (Batchelor 1967) is used to explain different patterns of seismicity during the slab sinking. Also we use two-dimensional numerical model of a descending slab to study the stress distribution during the process of slab sinking and delamination.

2. Shear Stress Evolution Induced by the Descending Vrancea Slab – A Corner Flow Model

In this section we consider a simple fluid dynamical model which provides an explanation for the observed distribution of seismicity in the Vrancea region. The Vrancea slab, which is believed to sink beneath the SE-Carpathian arc, separates the mantle into two portions (or two corner sub-domains). Considering that stresses released in earthquakes are related to the level of shear stress, we calculate shear stress distributions in the each corner sub-domain by using an analytical model for corner flow (Batchelor 1967). McKenzie (1969) used the corner flow model to study subduction zone dynamics and causes of plate motions. Later Stevenson and Turner (1977) and Tovish et al. (1978) used the same model to investigate the torque balance on the slab and angle of subduction for Newtonian and non-Newtonian rheologies of the mantle.

The descending slab must induce stresses within the surrounding mantle. Its motion will therefore influence the flow of the mantle. The principal forces determining the motion are the normal forces on the upper and lower surfaces of the slab (due to pressure variations in the surrounding mantle), gravity and resistance forces. Although the latter forces can contribute to the estimation of shear stresses on the slab, we follow McKenzie (1969) and Tovish et al. (1978) and omit resistance forces from the consideration to analyse viscous stresses only. Body forces caused by lateral thermal (density) variations are neglected, because our analytical model is purely mechanical (no thermal effects are considered in this section). Another assumption of the model is that flow in the mantle is governed by a viscous constitutional relationship, although it was shown that the shear stresses on the slab are reduced insignificantly, if the mantle behaves as non-Newtonian fluid (Tovish et al. 1978). Thus several of these assumptions are unlikely to be valid for mantle flow, but they enable analytical solutions to be obtained.

The corner flows are assumed to be two-dimensional. Figure 3 illustrates how the descending slab divides the mantle into two corners where flows are induced by the motion of the slab. We locate the Vrancea region at the origin of co-ordinates $x_1=0$, $x_2=0$. Axes Ox_1 and Ox_2 are directed leftward and downward, respectively. Surfaces $x_2=0$, $x_1<0$ and $x_2=0$, $x_1 \geq 0$ move towards the trench ($x_1=0$) with constant velocity U_1 and $-U_2$, respectively. The descending slab extends from the origin of co-ordinates downward at the dip angle α to the positive x_1 axis. The slab moves with constant velocity U_3 and/or vertical velocity U_4 (due to gravity). The model slab divides the viscous flow into two corners: "Transylvanian basin" corner (TB corner) and "East-European craton" corner (EEC corner). The applied velocities induce a viscous flow, and the flow and tectonic shear stress are determined within the two corners.

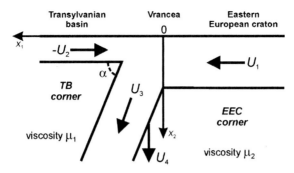

Fig. 3. Geometry and applied velocities of the model of corner flow induced by the descending slab beneath the Vrancea region.

The velocity components (v_1, v_2) of the mantle flow and maximum tectonic (deviatoric) shear stress τ_{max} in each corner can be found from the following expressions (Turcotte and Schubert 2002; Eqs. 6-110 and 6-111)

$$v_1 = -(A_i x_1 + B_i x_2)\frac{x_1}{x_1^2 + x_2^2} - B_i \arctan\frac{x_2}{x_1} - D_i, \tag{1}$$

$$v_2 = -(A_i x_1 + B_i x_2)\frac{x_2}{x_1^2 + x_2^2} + A_i \arctan\frac{x_2}{x_1} + C_i \tag{2}$$

$$\tau_{max} = 2\mu_i \frac{|A_i x_2 - B_i x_1|}{x_1^2 + x_2^2} \tag{3}$$

where A_i, B_i, C_i, and D_i are constants, and $i=1$ and $i=2$ correspond to TB and EEC corners, respectively. Their values are determined by boundary conditions.

The viscosity of cooled mantle material beneath the old EEC ($\mu_2=10^{21}$ Pa s) is assumed to be only five times higher than that of the mantle beneath the young TB ($\mu_1=2\times10^{20}$ Pa s). High mantle temperature and fluids beneath the TB may decrease the viscosity drastically, and hence the ratio between the mantle viscosities beneath the EEC and TB would be even larger.

The boundary conditions for the TB corner are $v_1 = -U_2$, $v_2 = 0$ at $x_2 = 0$, $x_1 > 0$ (or $\arctan(x_2/x_1) = 0$) and $v_1 = U_3 \cos\alpha$, $v_2 = U_3 \sin\alpha + U_4$ at $x_2 = x_1 \tan\alpha$ (or $\arctan(x_2/x_1) = \alpha$). An application of these boundary conditions to the equations for velocity leads to the following expressions for constants A_1, B_1, C_1, and D_1:

$$A_1 = \frac{-U_2\sin^2\alpha + U_3\alpha\sin\alpha + U_4(\alpha + \sin\alpha\cos\alpha)}{\alpha^2 - \sin^2\alpha} \qquad (4)$$

$$B_1 = \frac{-U_2(\alpha - \sin\alpha\cos\alpha) - U_3(\alpha\cos\alpha - \sin\alpha) + U_4\sin^2\alpha}{\alpha^2 - \sin^2\alpha} \qquad (5)$$

$$C_1 = 0, \qquad D_1 = -A_1 + U_2 \qquad (6)$$

The boundary conditions for the EEC corner are $v_1 = U_1$, $v_2 = 0$ at $x_2 = 0$, $x_1 < 0$ (or $\arctan(x_2/x_1)=\pi$) and $v_1 = U_3\cos\alpha$, $v_2 = U_3\sin\alpha + U_4$ at $x_2 = x_1\tan\alpha$ (or $\arctan(x_2/x_1)=\alpha$). Substituting the boundary conditions into the equations for velocity, we obtain the following expressions for constants A_2, B_2, C_2, and D_2:

$$A_2 = \frac{U_1\sin^2\alpha - U_3(\pi-\alpha)\sin\alpha - U_4(\pi-\alpha-\sin\alpha\cos\alpha)}{(\pi-\alpha)^2 - \sin^2\alpha} \qquad (7)$$

$$B_2 = \frac{-U_1(\pi-\alpha+\sin\alpha\cos\alpha) - U_3(\cos\alpha(\pi-\alpha)+\sin\alpha) + U_4\sin^2\alpha}{(\pi-\alpha)^2 - \sin^2\alpha} \qquad (8)$$

$$C_2 = -\pi A_2, \qquad D_2 = -A_2 - \pi B_2 - U_1 \qquad (9)$$

We consider three subsequent phases of the evolution of the descending lithosphere beneath the Vrancea: (i) active subduction ($\alpha=30°$, $U_1=U_3=5$ cm yr^{-1}); (ii) slab steepening due to gravity and slab roll-back ($\alpha=60°$, $U_2=U_4=5$ cm yr^{-1}); and (iii) gravity-driven slab sinking ($\alpha=85°$, $U_4=5$ cm yr^{-1}). Figure 4 shows the flow field and contours of constant shear stress for the model.

According to Sperner et al. (2001) the dip angle of the Vrancea subduction in Miocene times was about 25°. Fig. 4,*a* illustrates the pattern of mantle flow and distribution of stress during the active subduction of the Vrancea lithosphere. The pattern of the flow and shear stresses coincide with that of the McKenzie model (1969). Namely, the maximum shear stresses are concentrated at the upper surface of the descending slab. The model predictions are in good agreement with the locations of hypocentres of earthquakes within the so-called Wadati-Benioff zones associated with the active subduction in other seismic belts (e.g., Zhao 2001).

When the active subduction beneath the SE-Carpathians terminated, a rollback movement of the slab resulted in a redistribution of shear stress in the mantle. The maximum shear stresses are shifted from the upper to lower surface of the slab due to changes in the applied velocities (Fig.4,*b*). This effect is amplified during the final phase of slab evolution when the lithospheric plate sinks into the mantle driven only by gravity (Fig.4,*c*). According to Eq. (3), the higher the mantle viscosity beneath the EEC, the greater the magnitude of shear stress at the corner. The area of the maximum shear stresses in the model roughly coincides with the depth range of intermediate-depth events (IDE in Fig. 4) in SE-Carpathians. Ismail-Zadeh et al. (1999) showed that large earthquakes, generated by a dynamic model of a rigid slab descending in the viscoelastic mantle, are also concentrated at the lower surface of the Vrancea slab, but not at its upper surface.

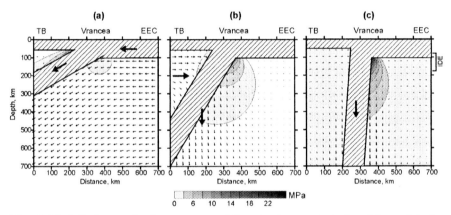

Fig. 4. Modelled tectonic shear stress and flow in the mantle induced by a slab descending beneath the Vrancea region for three phases of slab evolution: (a) active subduction, (b) slab steepening, and (c) slab sinking only due to gravity. Flow is indicated by the arrows, the shading represents the magnitudes of the shear stresses. IDE, intermediate-depth earthquakes.

Although our model presented here is based on simplified assumptions, it illustrates how changes in the dynamics of the descending slab result in a significant redistribution of shear stresses and hence in spatial changes of seismicity.

3. Past, Present and Future of the Vrancea Slab Inferred from Numerical Models

In this section we present two-dimensional numerical models of the slab evolution. We compute the flow and tectonic stresses induced by the descending slab during the processes of slab delamination and full detachment of the slab from the crust.

3.1 Model Description

In the numerical models the mantle is at rest before the onset of slab descend, thus convectively neutral. The motion is only caused by the descending slab. The initial geometry and boundary conditions for the models are shown in Fig. 5. A viscous incompressible fluid with variable density and viscosity fills the model region $0 \leq x \leq L$, $-H \leq z \leq h$ divided into five sub-domains by material interfaces: atmosphere above the surface, upper crust, lower crust, slab, and mantle. The density ρ and viscosity μ are constant within each, the upper and lower crust and mantle. The topography line approximates a free surface, because the density of the upper layer (the atmosphere) equals zero, and the viscosity is sufficiently low compared to that in the crust. The slab is modelled as being denser than the surrounding mantle, and therefore tends to sink gravitationally. The density and viscosity of the slab depend on the temperature T. Since we concentrate on an analysis of stresses induced by the sinking slab, the heat transfer in the mantle is neglected, although we understand its importance in general models of mantle convection.

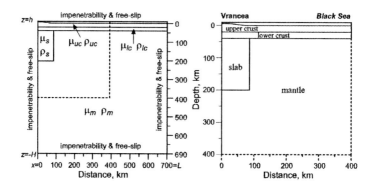

Fig. 5. Geometry of the model region with boundary conditions as indicated. The right panel presents an enlarged view of the model sub-region marked in the left panel by a dashed line.

We tested our models with respect to the stability of the results to variations of the density contrast, viscosity ratio between the slab and the surrounding mantle, and geometry of the slab. We considered (1) a density contrast ranging from 30 to 100 kg m^{-3}; (2) several values of the viscosity ratio: 10, 100, and 500, using a fixed density contrast of 70 kg m^{-3}; and (3) a variation of the initial depth of the slab penetration into the mantle ranging from 150 to 400 km. The model results show their robustness to these variations.

We solve the equation of motion (the Stokes equation) in terms of the stream function

$$4\frac{\partial^2}{\partial x \partial z}\mu(x,z)\frac{\partial^2 \psi}{\partial x \partial z}+\left(\frac{\partial^2}{\partial z^2}-\frac{\partial^2}{\partial x^2}\right)\mu(x,z)\left(\frac{\partial^2 \psi}{\partial z^2}-\frac{\partial^2 \psi}{\partial x^2}\right)=-g\frac{\partial \rho(x,z)}{\partial x} \qquad (10)$$

where $x=x_1$ and $z=x_2$. We assume impenetrability (no flow out of and into the model region) and free-slip boundary conditions, considering external forces to be negligible:

$$\psi = \partial^2\psi/\partial x^2 = 0 \quad \text{at } x=0 \text{ and } x=L, \qquad (11)$$

$$\psi = \partial^2\psi/\partial z^2 = 0 \quad \text{at } z=-H \text{ and } z=h.$$

Temperature within the slab is calculated from the following equation (McKenzie 1969):

$$\frac{T(x,z)}{T^*}=1-\frac{2}{\pi}\exp\left[\left(-\text{Re}+\left[\text{Re}^2+\pi^2\right]^{1/2}\right)\frac{z}{l}\right]\sin\frac{\pi(l-x)}{2l} \qquad (12)$$

where T^* is the temperature of the mantle surrounding the slab, $\text{Re}=\frac{\rho_s c_p l}{2\kappa}\frac{\partial \psi}{\partial x}$, ρ_s is the initial density of the slab, c_p is the specific heat at constant pressure, l is the slab thickness, and κ is the thermal conductivity of the slab.

The temperature-dependent density and viscosity of the slab are found from the following equations:

$$\rho(T)=\rho_s\left[1-\alpha_T(T-T_0)\right], \qquad \mu(T)=\mu_s\exp\left(\frac{E}{RT}-\frac{E}{RT_0}\right), \qquad (13)$$

where μ_s is the initial viscosity of the slab, T_0 is temperature at the bottom of the crust, α_T is volumetric coefficient of thermal expansion, E is the activation energy, and R is the universal gas constant. The positions of the material interfaces as functions of time are governed by the following differential equations:

$$\frac{dX}{dt}=\frac{\partial \psi}{\partial z}, \qquad \frac{dZ}{dt}=-\frac{\partial \psi}{\partial x}, \qquad (14)$$

where the points (X, Z) are on the initial interfaces at $t=0$. The initial distributions $(t=0)$ of density and viscosity and the positions of the material interfaces are known.

The maximum tectonic shear stress τ_{max} is given by

$$\tau_{max} = \left[\frac{1}{2}\left(\tau_{xx}^2 + \tau_{zz}^2 + 2\tau_{xz}^2\right)\right]^{1/2} = \mu\left[4\left(\frac{\partial^2\psi}{\partial x\partial z}\right)^2 + \left(\frac{\partial^2\psi}{\partial z^2} - \frac{\partial^2\psi}{\partial x^2}\right)^2\right]^{1/2}, \qquad (15)$$

where τ_{ij} $(i, j = x, z)$ are the components of the tectonic (deviatoric) stress tensor.

To solve the equations, that is, to compute the dependence of velocity, slab temperature, material interfaces, and shear stress on time, we employ the Galerkin method and finite element codes developed by Naimark et al. (1998). The model region is divided into rectangular 98×94 elements in the x and z directions.

We use dimensionless variables, whereas in presenting the results for stress and velocity we scale them as follows: the time scale t^*, the velocity scale v^*, and the stress scale σ^* are taken respectively as $t^* = \mu^*/[\rho^* g(H+h)]$, $v^* = \rho^* g(H+h)^2/\mu^*$, and $\sigma^* = \rho^* g(H+h)$, where μ^* and ρ^* are the typical values of mantle viscosity and density. The parameter values used in the modelling are listed in Table 1.

Table 1. Model parameters

Notation	Meaning	Value
c_p	specific heat at constant pressure, J kg^{-1} K^{-1}	1000 (*1*)
E	activation energy per mole, kJ mol^{-1}	2 (*1*)
g	acceleration due to gravity, m s^{-2}	9.8
h	height above the surface, km	5
$H+h$	vertical size of the model, km	700
l	thickness of the slab, km	90
L	horizontal size of the model, km	700
R	universal gas constant, J mol^{-1} K^{-1}	8.3 (*1*)
t^*	time scale, yr	14
v^*	velocity scale, m yr^{-1}	5×10^4
T_0	temperature at the bottom of the lower crust, K	873 (2)
T^*	temperature of the mantle, K	1573 (2)
α_T	volumetric coefficient of thermal expansion, K^{-1}	3×10^{-5} (*1*)
κ	coefficient of thermal conductivity, W m^{-1} K^{-1}	4 (*1*)
μ^*	typical value of viscosity, Pa s	10^{19}
μ_{uc}	effective viscosity of the upper crust, Pa s	3×10^{20}
μ_{lc}	viscosity of the lower crust, Pa s	5×10^{19}
μ_m	viscosity of the mantle, Pa s	3×10^{20}

Notation	Meaning	Value
μ_s	initial viscosity of the slab, Pa s	10^{21}
ρ^*	typical value of density, kg m^{-3}	3.37×10^3
ρ_{uc}	density of the upper crust, kg m^{-3}	2.76×10^3
ρ_{lc}	density of the lower crust, kg m^{-3}	2.97×10^3
ρ_m	density of the mantle, kg m^{-3}	3.3×10^3
ρ_s	initial density of the slab, kg m^{-3}	3.37×10^3
σ^*	stress scale, Pa	2.36×10^{10}

(1) Turcotte and Schubert 2002; *(2)* Demetrescu and Andreescu 1994

The viscosity being a least-known physical parameter is the only tuning parameter in our numerical models. We choose the value of typical viscosity μ^*, entering into the scaling relationships for t^* and v^*, so that the times of descending slab evolution predicted by the models are close to realistic geological times. The viscosity of the model mantle is chosen to be in agreement with the average viscosity for the upper mantle (long-dashed curve in Fig. 2, Forte and Mitrovica 2001). The densities of the upper and lower crust in the model are the averaged densities converted from P-wave velocities (Hauser et al. 2001).

3.2 Evolution of the Descending Slab from the Late Miocene to the Present

In the numerical model we assume that the initial thickness of the crust is 40 km, and the slab penetrates into the mantle to the depth of 200 km. The position of the model slab approximates the location of the slab at the end of Miocene times. Figure 6 presents the model evolution of the descending slab from 6.7 Ma ago to the present day. To enhance a visualisation of the numerical results, we present a portion of the model limited to the depth and width of 400 km.

At the early stage of the descending slab evolution (6.7 Ma ago) maximum shear stresses are high in the upper portion of the slab (~30 MPa) and in the upper crust (~20 MPa) and low (~10 MPa) in the lower crust and mantle (Fig. 6a, left panel). The stress decreases within the crust during the time between 6.7 Ma and 3.5 Ma, while it increases within the upper portion of the slab (Fig. 6b). Until present the model slab penetrates the mantle to the depth of about 350 km, and high maximum tectonic shear stresses (~35 to 55 MPa) are predicted at a depth range of about 80 to 180 km (Fig. 6c).

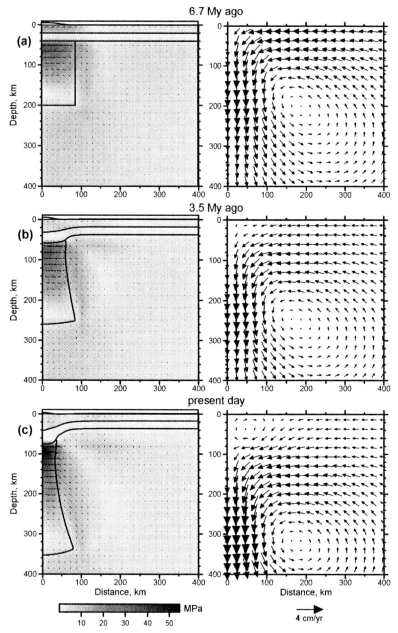

Fig. 6. Maximum tectonic shear stresses and axes of compression (left panel) and flow field (right panel) for the model evolution of the Vrancea slab (attached to the crust) descending in the mantle at successive times indicated in (*a*)-(*c*).

We consider this stage of the slab evolution to be related to the present-day situation in Vrancea, because (i) large seismic events are located at the same depths where our model predicts high shear stresses and (ii) a high velocity body beneath the Vrancea region is observed down to depths of 350 km (Wortel and Spakman 2000; Martin et al. 2001). A zone of low shear stress in the modelled lower crust (<10 MPa) is associated with the zone of low seismicity and observed zone of low seismic wave velocity shallower than 55 km (Hauser et al. 2001). Shear stresses decrease with depth within the slab as a result of temperature increase with depth.

Figure 6 (left panel) illustrates also the axes of maximum deviatoric compression. The axes of tension are perpendicular to the axes of compression. The axes of compression are subhorizontal in the upper portion of the slab (Fig. 6c) being in a good agreement with observations. Using numerous fault-plane solutions for intermediate-depth shocks, Oncescu and Trifu (1987) show that the compressional axes are almost horizontal and directed NW-SE.

The mantle flow induced by the descending slab is presented in Fig. 6 (right panel). Initially the motion of the crustal and uppermost mantle material is directed downward. The mantle flow is driven continuously by the descending slab (maximum rate of slab descent in Fig. 6c is about 4 cm yr^{-1}), whereas the upper crust subducted to the depths of about 40 km starts to rebound isostatically (uplift rate in the crust is about 0.5 cm yr^{-1}).

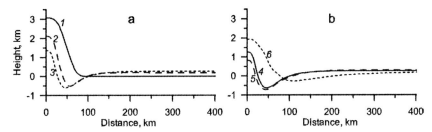

Fig. 7. Surface topography predicted by the numerical models of (a) attached and (b) detached Vrancea slab: Curves: *1*, 6.7 Ma ago (initial position of the model surface topography); *2*, 3.5 Ma ago; *3*, at present-day; *4*, in 0.5 Ma after slab detachment; *5*, in 2.0 Ma; and *6*, in 8.5 Ma.

The shape of the slab is controlled by the circulation of mantle material. The slab becomes thinner at the shallow levels (100-140 km) and thicker below (300-350 km), while the crust thickens above the slab due to flow of the crustal material induced by the descending slab. The motion results in changes of the surface topography contributing to the evolution of Carpathians and foredeep basin development. (Fig. 7a). An initial maximum elevation of the Carpathian mountains was prescribed to be about 3 km (curve *1*, Fig. *7a*). As the Vrancea slab sinks (curve *2* and *3*), a significant basin developed in the foredeep area and the maximum height of the Carpathians decreased to the today observed level of ca. 1.5 km (curve *3*).

3.3 Evolution of the Descending Slab After Slab Break-Off

To study the evolution of shear stress after slab break-off, we develop a model of the descending slab detached from the crust. The initial configuration of the slab and the crust were taken from the previous model.

Meissner and Mooney (1998) estimated the depth to potential decoupling zones between the descending lithospheric slab and crust by calculating lithospheric viscosity-depth curves based on reasonable geotherms and models of lithospheric composition. They found that zones of reduced viscosity are located within the lower crust and several tens of kilometres below the Moho. In this model we replace the lower crustal material brought to depths of 60 to 80 km due to slab pull by the mantle material. With this assumption, we introduce a break between the descending slab and crust. Instead of modelling the process of slab break-off itself, we concentrate on study of the stress evolution after slab detachment.

Figure 8 presents the model evolution of the detached slab for the next 8.5 Ma. Maximum shear stresses are still sufficiently high (~50 MPa) in the upper portion of the slab in 0.5 Ma after slab detachment (Fig. *8a*, left panel). Moreover, the stress increases at depths about 60-80 km because of the replacement of less viscous lower crustal material by more viscous mantle material. The slab continues to pull the detached crust downwards, while it continues to sink to depths of about 500 km (Fig. *8b*). Necking of the slab develops in its upper portion (at depths of 150 to 400 km), and around this narrow area the shear stresses are reduced. After about 8 Ma the slab reaches the lower boundary of the model which corresponds to the boundary between the upper and lower mantle (Fig. *8c*). Maximum shear stresses in the slab drops to its lowest level (~10 MPa), and the slab can no longer control the dynamics of the overlaying crust and uppermost mantle.

The mantle flow induced by the descent of the detached slab is presented in Fig. 8 (right panel). In 2 Ma after the slab detachment the flow is divided into two cells. The shallow cell above the slab involves the crust and uppermost mantle material in the circulation. The flow in the cell (with the rate of about 1 cm yr^{-1}) is associated with the isostatic rebound of the subducted crust. The deeper cell in the mantle is induced by the sinking slab, and its flow rate is about 5 cm yr^{-1}.

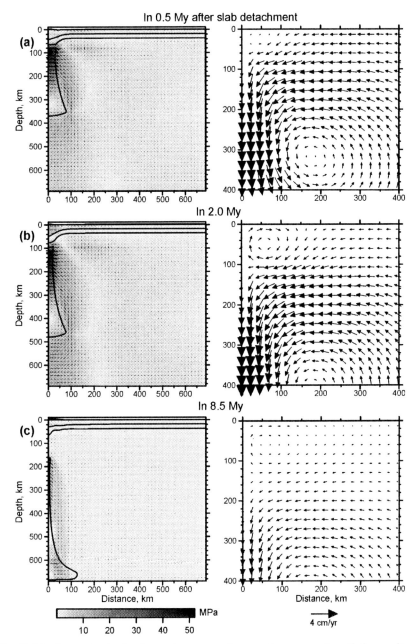

Fig. 8. Maximum tectonic shear stresses and axes of compression (left panel) and flow field (right panel) for the model evolution of the Vrancea slab (detached from the crust) descending in the mantle at successive times indicated (a)-(c).

Fig. 7*b* shows that the Carpathians subside even after slab detachment (curve *5*), although the subsidence slows down. In the model the rise of the Carpahians began in 2 Ma after the slab detachment (curve *6*), but the upward movements in the crust and uppermost mantle started earlier (before the slab detachment, see Fig. 6*c*). The temporal shift between the surface uplift and upward movements in the crust can be attributed to a balance between forces, which pull the mountains down, and forces that push the crust upward due to isostasy.

Recent GPS studies on vertical movements in the Vrancea region revealed a relative uplift of the crust (Dinter et al. 2001). If the model prediction on the onset of the surface uplift is correct, it should be expected that the Vrancea slab is already detached from the crust.

4. Discussion and Conclusions

There are essential distinctions between seismicity patterns of active subduction zones and those observed in continental collision zones. The subduction zone surrounding the Pacific Ocean is an extended structure several thousands of km in length and a few hundreds of km in width. Earthquakes with focal depths up to 60 km dominate these regions. These events are associated with subducting lithospheric slabs of mostly shallow subduction angles, and their hypocentres are located at or near upper surfaces of the slabs. In contrast to the concentration of epicentres along this subduction zone, the seismicity of the continental collision zones (e.g. Alpine-Himalayan orogenic belt) is diffuse, reflecting the greater width of the deformation pattern of this type of plate boundary.

Intermediate-depth seismicity in the SE-Carpathians (Vrancea region) is concentrated near and along the south-eastern margin of the high-velocity body (Fig. 2). The margin is considered to be a lower surface of the slab during its active subduction. The location of earthquakes is obviously not compatible with Wadati-Benioff zones of subducting lithosphere where earthquakes are concentrated near the upper side of the slab.

Using an analytical model for corner flow, we have showed here that the pattern of tectonic stress, induced by a descending slab in active subduction zones, differs from that in passive subduction zones. Maximum shear stress migrates from the upper surface of the descending slab to its lower surface due to changes in dynamics of the descending slab (from active subduction to sinking due to gravity only). Hence we conclude that the seismicity pattern of the final stage of a descending lithospheric plate differs completely from the pattern familiar from Wadati-Benioff zones.

Intermediate-depth events observed in several places in the world (the Mediterranean region, Carpathians, Caucasus, Zagros, Pamir-Hindu Kush, and Assam; Ismail-Zadeh et al. 2000) are associated with plate collisions. High-resolution seismic tomography in the regions of intermediate-depth seismicity is crucial to answer the question: whether the seismic events in these regions are generally located at the lower surface of descending slabs or whether this is a unique feature of the Vrancea seismicity.

The finite-element thermomechanical model of a descending slab allows us to explain the seismic activity in Vrancea on the basis of analysis of shear stress: the axes of compression are close to the horizontal as it is observed; and the maximum tectonic shear stress is found to be at depth of about 80 km to 180 km. The area of high maximum shear stress, predicted by the model, correlates with the region of Vrancea intermediate-depth events. The region of low seismicity at depths of 50 to 70 km is associated with a low viscosity zone in the model. An origin of the low viscosity zone might be due to either high temperatures at the relevant depths (Demetrescu and Andreescu 1994) or the lower crustal rocks brought down to the uppermost mantle depths during the slab descent.

An uplift of the crust begins before the initiation of slab detachment. The fate of the Vrancea slab is to be fully delaminated from the crust. The regional tectonic stresses are greatly reduced after following about 6 Ma slab detachment. The uplift and extension in the region predicted by the model are in a good agreement with observations in regions of slab detachment (e.g. Central Apennines, Wortel and Spakman 2000). The results of our analytical and numerical models together with seismic and geodetic observations allow interpreting the present-day dynamics of the Vrancea lithosphere as an ongoing process of detachment of the oceanic lithosphere from the overlaying crust.

Acknowledgements

We are very grateful to Karl Fuchs who invited us to present the paper at the Symposium "Challenges for Earth Sciences in the 21st Century". We thank W. Jacoby and J. Ritter for their constructive reviews of the manuscript. We are thankful to B. Sperner, F. Hauser, and M. Martin for useful discussions. Also we thank A. Wüstefeld for drawing of Fig. 2. This publication was supported by the Alexander von Humboldt Foundation and the Heidelberg Academy of Sciences and carried out under the auspices of the Collaborative Research Centre "Strong Earthquakes" (SFB 461).

References

Batchelor GK (1967) An introduction to fluid dynamics. Cambridge University Press, New York

Bijwaard H, Spakman W (2000) Non-linear global P-wave tomography by iterated linearized inversion. Geophys J Int 141: 71-82

Demetrescu C, Andreescu M (1994) On the thermal regime of some tectonic units in a continental collision environment in Romania. Tectonophysics 230: 265-276

Dinter G, Nutto M, Schmitt G, Schmidt U, Ghitau D, Marcu C (2001) Three-dimensional deformation analysis with respect to plate kinematics in Romania. Reports on Geodesy, 2: 1-21

Forte AM, Mitrovica JX (2001) Deep-mantle high-viscosity flow and thermo-chemical structures inferred from seismic and geodynamic data. Nature 410: 1049-1056

Fuchs K, Bonjer K, Bock G, Cornea I, Radu C, Enescu D, Jianu D, Nourescu A, Merkler G, Moldoveanu T, Tudorache G (1979) The Romanian earthquake of March 4, 1977. II. Aftershocks and migration of seismic activity. Tectonophysics 53: 225-247

Gibracea R, Frisch W (1998) Slab in the wrong place: Lower lithospheric mantle delamination in the last stage of the Eastern Carpathian subduction retreat. Geology 26(7): 611-614

Gutenberg B, Richter CF (1954) Seismicity of the Earth. Princeton University Press, Princeton, N.J., 2nd ed

Hauser F, Raileanu V, Fielitz W, Bala A, Prodehl C, Polonic G, Schulze A (2001) VRANCEA99 – the crustal structure beneath the southeastern Carpathians and the Moesian Platform from a seismic refraction profile in Romania. Tectonophysics 340: 233-256

Ismail-Zadeh AT, Keilis-Borok VI, Soloviev AA (1999) Numerical modelling of earthquake flows in the southeastern Carpathians (Vrancea): Effect of a sinking slab. Phys Earth Planet Inter 111: 267-274

Ismail-Zadeh AT, Panza GF, Naimark BM (2000) Stress in the descending relic slab beneath the Vrancea region, Romania. Pure Appl Geophys 157: 111-130

Linzer H-G (1996) Kinematics of retreating subduction along the Carpathian arc, Romania. Geology 24: 167-170

Martin M, Achauer U, Kissling E, Mocanu V, Musacchio G, Radulian M, Wenzel F, Calixto working group (2001) First results from the tomographic experiment CALIXTO '99 in Romania. Geophys Res Abst 3: 84

McKenzie DP (1969) Speculations on the consequences and causes of plate motions. Geophys J R Astron Soc 18: 1-32

McKenzie DP (1972) Active tectonics of the Mediterranean region. Geophys J R Astron Soc 30: 109-185

Meissner R, Mooney W (1998) Weakness of the lower continental crust: a condition for delamination, uplift, and escape. Tectonophysics 296: 47-60

Naimark BM, Ismail-Zadeh AT, Jacoby WR (1998) Numerical approach to problems of gravitational instability of geostructures with advected material boundaries. Geophys J Int 134: 473-483

Oncescu MC (1984) Deep structure of the Vrancea region, Romania, inferred from simultaneous inversion for hypocentres and 3-D velocity structure. Ann Geophys 2: 23-28

Oncescu MC, Bonjer KP (1997) A note on the depth recurrence and strain release of large Vrancea earthquakes. Tectonophysics 272: 291-302

Oncescu MC, Trifu CI (1987) Depth variation of moment tensor principal axes in Vrancea (Romania) seismic region. Ann Geophys 5: 149-154

Sperner B, Lorenz F, Bonjer K, Hettel S, Müller B, Wenzel F (2001) Slab break-off – abrupt cut or gradual detachment? New insights from the Vrancea region (SE Carpathians, Romania). Terra Nova 13: 172-179

Stevenson DJ, Turner JS (1977) Angle of subduction. Nature 270: 334-336

Tovish A, Schubert G, Luyedyk BP (1978) Mantle flow pressure and the angle of subduction: non-Newtonian corner flow. J Geophys Res 83: 5892-5898

Turcotte DL, Schubert G (2002) Geodynamics. Cambridge University Press, Cambridge, 2nd ed

Wenzel F, Achauer U, Enescu D et al.(1998a) Detailed look at final stage of plate break-off is target of study in Romania. EOS Transactions AGU 79: 589-594

Wenzel F, Lorenz FP, Sperner B, Oncescu MC (1998b) Seismotectonics of the Romanian Vrancea area. In: Wenzel F, Lungu D, Novak O (eds) Vrancea Earthquakes: Tectonics, Hazard and Risk Mitigation. Kluwer, Dordrecht, pp 15-25

Wortel MJR, Spakman W (2000) Subduction and slab detachment in the Mediterranean-Carpathian region. Science 290: 1910-1917

Zhao D. (2001) Seismological structure of subduction zones and its implications for arc magmatism and dynamics. Phys Earth Planet Inter 127: 197-214

List of Reviewers

Professor U. Achauer
EOST Strasbourg, 5 Rue René Déscartes, 67084 Strasbourg, France,
ulrich.achauer@eost.u-strasbg.fr

Professor J. Ansorge
Institut für Geophysik, ETH Hönggerberg, 8093 Zürich, Switzerland,
ansorge@tomo.ig.erdw.ethz.ch , or ansorge@bluewin.ch

Professor Z. Ben-Avraham
Department of Geophysics & Planetary Sciences, Dead Sea Research
Center, Tel Aviv University, Tel Aviv 69978, Israel, zvi@terra.tau.ac.il

Professor S. Cloetingh
Dept. of Sedimentary Geology, Vrije Universiteit, De Boelelaan 1085,
1081 HV Amsterdam, The Netherlands, cloeting@geo.vu.nl

Prof. H. Drewes
Deutsches Geodätisches Forschungsinstitut, Theoretische Geodäsie,
Marstallplatz 8, 80539 München, Germany,
drewes@dgfi.badw-muenchen.de

Prof.em. Dr. J. Eibl
Käthe-Kollwitz-Str. 26, 76227 Karlsruhe, Germany, eibl.josef@t-online.de

Professor M. Erdik
Bogazici University, Kandilli Observatory & Earthquake Research
Institute, Dept. Earthquake Engineering, Cengelkoy, 81220 Istanbul,
Turkey, erdik@boun.edu.tr

Professor H. Igel
Institut für Angewandte und Allgemeine Geophysik, Ludwig-Maximilian
Universität, Theresienstrasse 41, 80333 München, Germany,
igel@geophysik.uni-muenchen.de

Printed in the United Kingdom
by Lightning Source UK Ltd.
119883UK00007B/122